NO
APPARENT
DANGER

NO APPARENT DANGER

**THE TRUE STORY OF VOLCANIC DISASTER
AT GALERAS AND NEVADO DEL RUIZ**

VICTORIA BRUCE

HarperCollins*Publishers*

HarperCollins books may be purchased for educational, business, or sales promotional use. For information please write: Special Markets Department, HarperCollins Publishers, Inc., 10 East 53rd Street, New York, NY 10022.

FIRST EDITION

Designed by The Book Design Group

Cover photo by Norman Banks

Library of Congress Cataloging-in-Publication Data
Bruce, Victoria.
    No apparent danger : the true story of volcanic disaster at Galeras and Nevado del Ruiz / Victoria Bruce.—1st ed.
        p.   cm.
    Includes index.
    ISBN 0-06-019920-2
        1. Nevado del Ruiz (Colombia)—Eruption, 1985.   2. Lahars—Colombia—Nevado del Ruiz Region.   3. Galeras Volcano (Colombia)—Eruption, 1993. 4. Volcanism—Colombia—History—20th century.   I. Title.

QE523.N48 B78   2001
363.34'95'098613—dc21

                                                                00-054114

 00 01 02 03 04 10 9 8 7 6 5 4 3 2 1

CONTENTS

THE SCIENTISTS

VOLCANOLOGISTS

Norm Banks	U.S. Geological Survey	Washington, U.S.A.
Pete Hall	Escuela Politecnica Nacional	Quito, Ecuador
Patty Mothes	Escuela Politecnica Nacional	Quito, Ecuador
John Tomblin	United Nations	Switzerland

GEOLOGISTS

Marta Calvache	Ingeominas*	Pasto, Colombia
Dave Harlow	U.S. Geological Survey	California, U.S.A.
Darrel Herd	U.S. Geological Survey	Virginia, U.S.A.

SEISMOLOGISTS

Bernard Chouet	U.S. Geological Survey	California, U.S.A.
Fernando Gil	Ingeominas	Manizales, Colombia
Diego Gómez	Ingeominas	Pasto, Colombia
Bruno Martinelli	Swiss Disaster Relief Organization	Switzerland

* Colombian National Institute of Geology and Mines

| Fernando Muñoz | Ingeominas | Pasto and Manizales, Colombia |
| Roberto Torres | Ingeominas | Pasto, Colombia |

CHEMISTS

Andy Adams	Los Alamos National Lab	New Mexico, U.S.A.
Mike Conway	Florida International University	Michigan, U.S.A.
Fabio García	Ingeominas	Bogotá, Colombia
Nestor García	University of Caldas	Manizales, Colombia
Luis LeMarie	Escuela Politecnica Nacional	Quito, Ecuador
Andrew Macfarlane	Florida International University	Florida, U.S.A.
Igor Menyailov	Institute of Volcanology	Kamchatka, Russia
Alfredo Roldán	Instituto Nacional de Electrificación	Guatemala City, Guatemala
Stanley Williams	Arizona State University	Arizona, U.S.A.
José Arles Zapata	Ingeominas	Pasto, Colombia

GEOPHYSICISTS

Geoff Brown	The Open University	United Kingdom
Fernando Cuenca	Ingeominas	Bogotá, Colombia
Milton Ordoñez	Ingeominas	Pasto, Colombia

CIVIL ENGINEERS

| Bernardo Salazar | CHEC† | Manizales, Colombia |
| Carlos Trujillo | Universidad de Nariño | Pasto, Colombia |

† Central Hydroelectric Company of Caldas

NO
APPARENT
DANGER

PROLOGUE

IT'S LATE MORNING, 40 degrees with a strong wind, and we are standing on the summit of Galeras, an ample 14,000-foot volcano in southern Colombia. In my backpack I have two candy bars—offerings to appease the mountain, brought along at the urging of Alfredo Roldán, my Guatemalan guide. Alfredo has survived an eruption of Galeras once, and he isn't taking any chances.

Galeras doesn't like strangers.

The summit of Galeras is 500 feet above its crater. Below us, I can just make out four white specks moving against the 450-foot-high pile of rubble that is the volcano's young cone: white hard hats, required safety equipment for scientists working near the crater.

One of the four is Gustavo Garzón, a geologist from the volcano observatory in Manizales, Colombia, who is guiding three German scientists. Two nights ago, Gustavo, Alfredo, and I were graciously hosted by a local civil engineer with ties to the volcano observatory. There were two bottles of vodka, a boiling cauldron of cheese fondue, and what seemed like at least six packs of cigarettes. We spent the evening talking about the volcanoes we had worked on: Nevado del Ruiz and Galeras in Colombia, Pacaya in Guatemala, Mount Rainier in North America, Bezymianny in Russia. Alfredo brought out pictures of Pacaya erupting.

As is the case in most scientific circles, there is a closeness among the geologists here in Colombia—a nobody-understands-us-but-us camaraderie. We typically get lost in the most boring minutiae of our science, tell stupid jokes about gneiss and schist, become fascinated by the tiniest mineral grain or ripple marks across a chunk of sandstone, argue about the chemical makeup of the earth's core.

I am here atop this volcano because there is a story inside this barren landscape that links scientists and nonscientists. A series of events that tore apart a city, divided journalists and politicians, played the scientists against the people. A story with centuries-old roots that escalated over a decade and culminated on this very spot in 1993 in the deaths of nine people—Galeras's first and only victims in recorded time.

A man from the volcano observatory in Pasto stands on the rim next to us. He hands me a radio. Gustavo is on the other end, a half-mile away—he is the size of a gnat from here. He's standing next to a fountain of steam pouring from a hole in the side of Galeras's cone.

"Nice day, isn't it?"

"Yeah, great," I say through chattering teeth.

I hand the radio back, and we hear Gustavo crackling through. Several kids have climbed down below, and Gustavo is yelling at them to leave. It's dangerous. Clouds as thick as cappuccino foam pour in and out of the crater. I wonder how the kids got down there. Galeras's crater is officially off-limits.

We sit on rocks near a rope that is cleated to the summit, and we watch Galeras breathe, wispy white plumes flowing vigorously from its crater. Ten minutes later, the rope begins to flutter; a tiny form appears, wearing flat-soled rubber boots that slide and scrape on the loose rock of the precipitous slope. A progression of boys follows.

The youngest looks about 8 years old. Their faces are round and smooth and copper brown. They are dressed in baggy trousers and colorful sweaters. Some are missing front teeth. They have no helmets or safety gear. They wave, smiling, and call to me in broken English.

Following the boys up the rope are the fathers, their faces carved and creased. They too are without hard hats, and they are dressed just like the boys. The men are construction workers repairing the police station. So this is how they gained access to the volcano.

I lift my head in the direction of a decrepit concrete structure a hundred yards away, a building that was nearly obliterated by the volcano in an eruption several months after that fateful day in 1993.

"You're rebuilding the police station?" I ask one of the workers, a man with an angled face, a knit cap. He nods.

"Do you think Galeras will erupt again?"

He flashes a gapped-tooth grin and shrugs his shoulders. "No . . . yes . . . maybe."

He's not worried. To the people of the countryside, or campesinos, Galeras is family. The volcano may not like strangers, but he would never hurt his own.

Seven years ago, Alfredo was here with a team of scientists from Los Alamos. This time he is working for me. He is my guide, and I am asking him to relive the day of January 14, 1993. It looked much different then, he tells me. The crater is wider now. Deformes fumaroles are pouring much more profoundly, there is more steam coming from the volcano's throat. This ridge was much wider. He stops talking for a few seconds, then he remembers the moment that Galeras erupted, at 1:43 P.M. He points: "Here's where I was. Here's where the journalists were. Here's where the rocks were falling." He becomes quiet and lights a cigarette with his small magnifying glass and the sun, which has momentarily sprung from the cloud cover.

Gustavo radios in to Bruno Martinelli, who is keeping tabs on the seismometers at the observatory 5 miles away in the city of Pasto. The man next to us with the radio is acting as a link between

Bruno and Gustavo's team working in the crater, so we can hear both sides of the conversation.

Gustavo: "The gas temperatures are coming in at a thousand degrees Celsius."

Bruno: "No way. It's not possible. Something is wrong with the equipment." I can't see from here, but Alfredo describes how they are taking the temperature of the superhot steam with long wires linked to a thermocouple. A thousand degrees Celsius? I don't believe it either. Sounds way too hot.

I walk east along the ridge and stand by myself, looking down into the carved-out caldera. I am 33 years old, the same age that Marta Calvache was on that day in January 1993. Like Marta, I am a geologist, but Marta grew up at the back door of this volcano, on a small farm in the village of Consacá, and I was raised in the sub-urbs of southern California. While I hemmed and hawed and won-dered what to do with my life, Marta literally took on mountains. I would like to imagine that I could follow in her footsteps, but I'm honestly not sure I could. Marta is a hero. That January day, Gale-ras's crater roared like a jet engine, shooting out black clouds with roots of incandescent fire. While a dozen men stood frozen in fear on the summit, Marta descended into the inferno.

I picture the 5-foot 1-inch scientist climbing down the rope. She tries to run across the rubble. Scorching rocks burn through her boots and sizzle on the cold ground. The volcano roars. Near a 3-foot boulder, she finds her professor, Stanley Williams. He is scorched and twisted and bloodied and is crying for help. Close to him are four more victims of the volcano, unrecognizably contorted, deep holes in their skulls. Their clothes have seared onto their lifeless bodies.

It takes over two hours for Calvache and three others to carry Williams to safety. Six more narrowly escape with their lives. Nine others would never return.

There are incredible heroes in this desolate cauldron, and there are ghosts. The path that brought them here stretches back a decade and reaches 300 miles north along the Cordillera Central. There, in

1984, Marta Calvache and a group of young Colombian scientists worked to uncover the deadly secrets of another volcano, called Nevado del Ruiz. The two volcanoes are inextricably linked—by geology, by legend, and by scientific failing. It's impossible to understand what happened here at Galeras without first going back to the terrible tragedy at Nevado del Ruiz.

I take a deep breath and stare hard into the volcano's steaming interior. Boulders cast firm shadows to the west, and I watch them shrink away with the rising sun. Too soon, a breaking wave of white mist comes up and over the rim of the volcano and spills into the crater. Galeras disappears from view. I wait, hoping for another glimpse, another moment, but it is not to be. The clouds have settled like a blanket over the volcano's cone. I clap my gloved hands to bring back the circulation and walk toward Alfredo. It's time to go back.

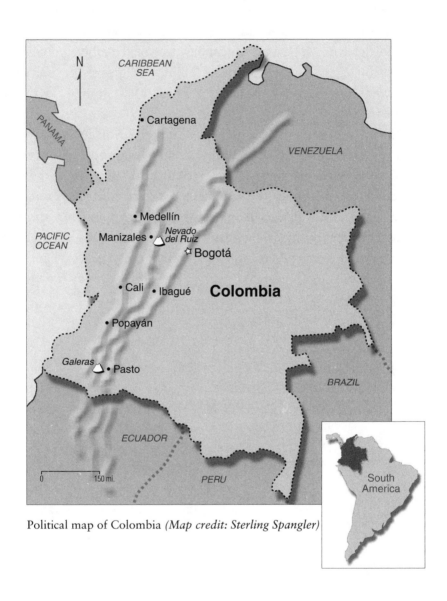

Political map of Colombia *(Map credit: Sterling Spangler)*

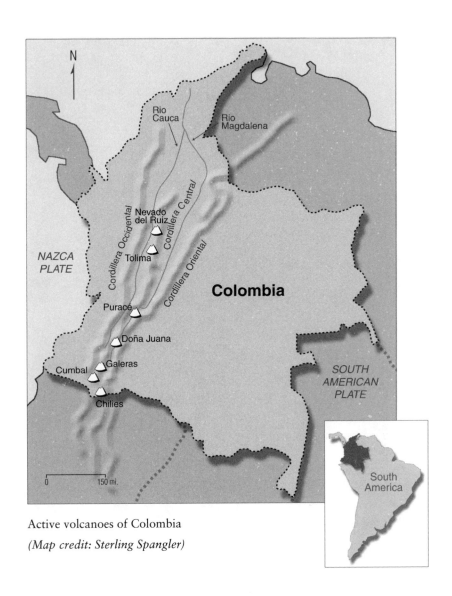

Active volcanoes of Colombia

(Map credit: Sterling Spangler)

PART I: NEVADO DEL RUIZ

CHAPTER 1: THE GREAT MOUNTAIN

ARBOLITO, COLOMBIA:
NOVEMBER 13, 1985

Bernardo Salazar and Fernando Gil, two young scientists working for the regional electric company, hiked up the steep ridge above the small farming village of Arbolito, Colombia. Above them, hidden by clouds, rose the sweeping summit of Nevado del Ruiz, a 17,454-foot ice-capped mountain and the highest active volcano in Colombia. Nestled within the northernmost reaches of the Andean Cordillera, the beautiful, tranquil Nevado del Ruiz symbolized security and protection to the people of the region.

Sleeping peacefully for as long as anyone could remember, the volcano had become restless during the previous year. The gentle rumblings and the tall column of steam flowing from its crater had frightened farmers living close to the mountain's flanks and unsettled coffee growers, politicians, and scientists in the cities below, but since there had been no sizable eruption, many people in the surrounding regions stopped taking the volcano seriously. "The great mountain considers us family," the locals said. "The volcano may get a little riled up, but it would never hurt its own."

Things had changed somewhat just two months before. On September 11, 1985, Nevado del Ruiz erupted. A thin coat of ash rained onto Manizales, the capital city of the western state of Caldas. Frightened residents swept ash from the narrow streets of their city and spoke about their mountain with the sort of concern normally reserved for an admired but volatile family matriarch.

The prosperous metropolitan center of Manizales sat high on a ridge, and scientists had assured residents that the city was not likely to be hurt by the volcano. What *did* worry scientists were the many thousands who lived in the valleys and low-lying plains surrounding Nevado del Ruiz. Yet even with such a specific threat, there was no money from the federal government to monitor the volcano. The data collected by scientists were slow to be analyzed or, worse yet, ignored altogether, making it almost impossible to accurately predict an eruption and take action before it was too late.

After the September eruption, however, Bernardo Salazar, a quiet, clean-cut civil engineer, had received a green light to add one more member to the team that had been working on Nevado del Ruiz since the beginning of the year. Salazar approached Fernando Gil, a bearded man with a booming laugh and a fellow alumnus of the University of Caldas.

"I was working as a civil engineer when Bernardo called to ask if I wanted to work on the volcano. I told him yes, of course I did," Gil says.

As Salazar explained to Gil by phone just days after the September eruption, they would visit all six of the seismic monitoring stations scattered around the volcano every day, by car and by foot, and gather the seismograph data. Gil would live with Salazar in Arbolito, just 3 miles northwest of the volcano's summit. They would start before dawn, Salazar explained, because the task of changing all the seismographs took the entire day. The roads that traveled around Nevado del Ruiz were cobbled with angular volcanic rocks or laid-in dirt and mud and carved by gullies. To get the broadest data possible, the seismic stations were scattered great distances over the volcano's massive cone.

The two scientists would then evaluate each of the carbon-coated records, cataloging the time and the duration of small scribbles that appeared on the seismographs. And every other week, Salazar would make the two-hour drive down the mountain and deliver the seismographs to the team's unofficial headquarters on the eleventh floor of the Banco Cafetero building in downtown Manizales. In the office space, which the scientists referred to as *Piso Once* (literally, the "eleventh floor"), Marta Calvache, a young geologist, and Nestor García, a chemical engineer, would then send the seismographs to Bogotá to be analyzed by government scientists from the National Institute of Geology and Mines.

That was the official routine—though, in fact, no results had come back to *Piso Once* for over two months. Even more discouraging, the hydroelectric company that had loaned Salazar the seismograph drums wanted their equipment back to use on other projects. Salazar felt like a year of hard work was quickly coming to an unsuccessful end.

Gil arrived in Arbolito just a week after receiving Salazar's invitation. He shared a two-room house with Salazar and two other housemates, Rafael Gonzales and Juan Duarte, technicians responsible for keeping the equipment working. The view from the front yard was the massive summit of Nevado del Ruiz, whose magnificent peak was often shrouded in thick clouds. When the clouds lifted, glaciers covering the top of the mountain looked like a menacing row of jagged shark's teeth. The ice, ribbed with gaping, sinuous crevasses, flowed from the volcano's broad summit down into steep valleys and ended in ominous pointed cusps. "It was amazingly beautiful, and I felt so fortunate," Gil says.

The year was 1985. Just over five years before, the tremendous eruption of Mount Saint Helens in Washington State rocked the scientific community. There were only a handful of scientists in the world who could accurately interpret seismic signals from an active volcano, and those specialists hadn't come to Nevado del Ruiz. "We were civil engineers and knew virtually nothing about volcanoes," says Gil. Despite their best intentions, Salazar and Gil had no idea how to interpret the small scribbles they were looking at, night after night, under the poor light of a single lamp in the cinder-block house.

The seismograph in Arbolito was placed high on a ridge. The hike up the steep, grassy slope had been part of Gil's daily commute for the past eight weeks, and now he didn't even need to stop and catch his breath on the way up. When Salazar and Gil reached the top, freezing winds whipped across the crest, and Salazar ducked into a makeshift plastic tent propped up by several crooked sticks. He negotiated a padlock that dangled from an oven-sized metal box. Under the heavy cover was the seismograph, a rotating drum 18 inches long and 8 inches in diameter. Salazar quickly lifted a needle that was suspended above the seismograph and removed the drum on whose face was etched a daylong history of the volcano's rumblings. The needle received its signals from a seismometer, a thermos-sized instrument planted beneath the ground on the high ridge. The seismometer detected activity—rocks fracturing, gases reaching very high pressures, sonic resonances—within Nevado del Ruiz. A volcanic explosion, even a very small one, sends shock waves into the surrounding rocks, and the seismograph needle registers the activity the way a polygraph monitors a nervous subject's vital signs.

It was midafternoon on November 13 when Gil and Salazar changed the last seismograph on the ridge above Arbolito. Salazar held the ends of the drum in his hands, twisting it so that he and Gil could see the entire record. A small scribble was etched into the carbon on the paper. Gil pointed out the sharp waves to Salazar.

"We looked toward the mountain but couldn't see anything because of the clouds," Gil says. "We hadn't heard any sound, so we figured that the signal was just noise—wind maybe." The signal had been recorded just over an hour before, and yet there had been no discernible sign that anything had happened on the volcano. The marking was bigger than what usually appeared on the seismographs, but Nevado del Ruiz appeared to be resting peacefully.

Cold and tired, his hands numb, Salazar packed away the first drum and took the second from Gil. This would record seismic activity within the volcano for the next twenty-four hours. He checked his watch and made a note in the margin: Arbolito, 13 November 1985, 4:30 P.M., Nevado del Ruiz.

A year earlier, in November 1984, mountain climbers on Nevado del Ruiz reported being rattled by earthquakes and seeing tall white billows of sulfur-laden gases rising from the summit crater. Worried that an eruption from the volcano could have catastrophic consequences to its business interests in the area, the electric company joined with the coffee growers in search of experts. What they found was that there was not a single scientist trained to deal with volcanic hazards in the entire country of Colombia. The regional electric company then corralled Bernardo Salazar and two more of its best young scientists, Marta Calvache and Nestor García, and gave them the job of keeping tabs on the restless volcano. Calvache was one of just two people in the entire country who had any understanding of a volcano's geology; García's job would be to test the acidity and temperature of hot steam escaping from the volcano's crater.

It was soon evident that Salazar, Calvache, and García, although quite dedicated and eager to explore the mountain, were completely untrained and ill-equipped to work on an active volcano by themselves. The result was a loud call to the United Nations for help. Ecuadorian, Swiss, U.S., and Italian scientists came to Manizales, but their often conflicting views about Nevado del Ruiz confused the locals, the politicians, and the young Colombian scientists. And there was no money for monitoring equipment or support from the Colombian government.

The political and geographical boundaries that divided the region made it even more difficult to deal with the threat of the volcano. The massive, complex mountain of Nevado del Ruiz, created by millions of years of lava flows, straddles three Colombian states, or departments: the Cafetera region, comprising Caldas and Risaralda to the west and Tolima to the east. The Cafetera region was a prosperous area that had escaped the political and economic tragedies that plagued most of Colombia. The relatively wealthy population of Caldas had taken many measures over the previous year to prepare for a possible disaster. The poorer and less educated population of Tolima had not.

Lately, however, the state of Tolima had started taking the volcano very seriously. Scientists from around the world predicted that many of the towns lying in the low, flat plains to the east of the volcano were in grave danger if Nevado del Ruiz erupted.

Twice previously, once in 1595 and again in 1845, eruptions on Nevado del Ruiz had melted the mountain's glaciers and sent a deadly avalanche of mud and debris to the place that was now home to over 30,000 people.

On October 7, Colombian scientists came to a terrifying conclusion that was reported on the front page of Bogotá's *El Espectador*: If Nevado del Ruiz erupted, deadly mudflows would almost certainly obliterate Tolima's low-lying city of Armero—and residents would have two hours to evacuate the city.

Gil and Salazar hadn't realized that the last seismometer they'd checked earlier in the afternoon on November 13 had left—in a language neither understood—a very clear message.

Nevado del Ruiz had already erupted.

The billowing cloud of ash and steam that exploded from the volcano at 3:09 P.M. rose several miles into the sky, but it was hidden from view on the cloudy afternoon. It wasn't until several hours later, when ash began to fall in northern Tolima along with heavy rain, that anyone knew the volcano had erupted. Civil Defense in Tolima was on alert, but so far there had been no calls for help.

In the quiet city of Armero, two dozen geology students from the University of Caldas watched as the rain became dirty with ash. They knew instantly what had happened. The group was in the middle of a two-day field trip to study the paleontology of the Tolima region. They'd gotten a late start from Manizales in the morning and were still a few hours from their final destination. It was already dark, and Jorge Dorado, the professor and leader of the field trip, decided that rather than go on to Ibagué, a few hours to the south, as they had planned, the group would spend the rainy night in Armero.

In the University of Caldas geology department's ongoing debate about whether or not Nevado del Ruiz was a great danger to the surrounding countryside, Dorado was among those who had openly scoffed at the notion. "He was sure the volcano wasn't dangerous," says Fernando Muñoz, who was a fellow professor at the university. But on this dark night, as the ashy rain fell, his students noticed that the usually boisterous professor had become very quiet.

For most of the students, though, the eruption was science in action, and they loved being blanketed with ash from the powerful Nevado del Ruiz volcano. They skipped through the wet streets of the town, letting the dirty rain soil their clothes and their outstretched palms.

But one of the students, Jorge Estrada, was anxious. Just a few weeks before, he'd been helping scientists from the National University in Manizales work on hazard maps of Nevado del Ruiz. The preliminary maps were made with colored pencils. Sectors that would be affected by pyroclastic flows were drawn in red, areas of ash fall in yellow, and mudflows in brown. "I remember Jorge Estrada taking a pencil to a map," says Muñoz. With the edge of the pencil lead, he drew a thick line that started at the summit of Nevado del Ruiz and followed the curves of the Rio Lagunillas. Then, with a dark brown marker, he obliterated the small city of Armero. "He was laughing, and he said, 'There you go, Armero. Now we're burying you in the mud.'"

In Tolima's capital city of Ibagué that evening, the Regional Emergency Committee of the state of Tolima told Civil Defense stations and the Red Cross to sound the alarm to the low-lying cities along the rivers *if necessary*—that is, if there were reports that the mudflows were coming. But by 7:30 that evening, the heavy rain in Armero had turned to a steamy drizzle, and the ash had quit falling.

The Regional Emergency Committee reported to the Red Cross of Tolima that things were quieting down.

On the church loudspeaker system that reached most of the small city, the priest of Armero said the rosary and told listeners that they had nothing to fear. El Ruiz was quiet again, he said. It was only a small eruption. Everything would be fine. *Put your faith in God.*

On the northwestern flank of Nevado del Ruiz, in the small house in Arbolito, Fernando Gil and Bernardo Salazar had gotten word by radio from Manizales that there was ash falling in Tolima. As a scientist, Gil was disappointed that he hadn't seen the eruption, but he was even more irritated that he hadn't been able to tell by looking at the seismograph data that the volcano had erupted. "It was incredible," he says. "We saw the signal, but we had no idea what it meant." What good were he and Salazar doing, he wondered, if they couldn't even tell what an eruption looked like on a seismograph? Gil was relieved when the following radio report stated that there were no injuries from the eruption, and Civil Defense in Tolima said things appeared quiet.

Gil spent some time working on the seismographs he and Bernardo Salazar had collected. He dipped the paper into a resin that would fix the carbon so that the patterns in the graphite wouldn't be destroyed and laid them over the kitchen chairs to dry.

After many miles of hiking over the steep slopes of the volcano, the young scientists were tired, and at 9 P.M. they prepared for bed. Strong wind rattled the windows in the cinder-block house. Gil turned off the lamp and drifted into sleep.

Suddenly, the ground shook beneath the house as a loud explosion crashed through the valley. Another blast followed just seconds later. The four men scrambled out of their beds and ran to the door. The sky was dark. The volcano exploded a third time, and the entire cloud above the mountain lit up like a tremendous lantern. Almost instantaneously, heat from the eruption created a thunderstorm over the volcano. The clouds flashed with lightning, followed by crackling thunder.

"I remember Bernardo took the radio and he called to CHEC [The Central Hydroelectric Company of Caldas]. He was yelling, 'The volcano erupted . . . Nevado del Ruiz erupted,' " Gil says.

At the electric company headquarters in Manizales, a napping guard awoke and grabbed the microphone on his desk. He wasn't sure what he'd just heard.

"Can you repeat that?" he called.

"Ruiz has *erupted*!" Salazar yelled into the radio.

"Is anyone injured? Do you need help?"

"Not here," Salazar shouted. "Alert Civil Defense about the eruption. Tell them to start the evacuations!"

Pebbles began to rain from the sky, and the men ducked back through the doorway of the house. The falling rocks sounded like machine-gun fire as they hit the tin roof. Larger pumice stones began to fall. A baseball-sized rock shattered the skylight, sending the men racing for cover.

Despite the hysteria, Gil was transfixed by the eruption. He had felt the explosion resonate throughout his body. "I felt like it was God reaching inside me, touching my soul," Gil says. The young scientist stepped into the night. He fell to his knees and bowed his head; humbled and awestruck, he began to cry.

Within minutes, the earth began to rumble again. The clamor grew louder and louder, the shaking and rumbling stronger and more ferocious every second. The four men stared at the river valley, a sharp V that came toward them directly from the mountain. "In the dark, we couldn't see anything, but we could hear it coming," says Gil. There was no mistaking the thunderous sound. The wall of mud moved furiously down the valley, growing louder as it flowed, ripping apart the soft earth it traveled through and becoming more and more deadly by the second.

His hand trembling, Salazar took the radio from his coat pocket and held it close to his mouth, yelling over the din of the approaching black avalanche.

"Manizales, the mudflows are coming. Alert the state of Tolima immediately!"

"Civil Defense is on alert," said the voice on the other end. "Things are under control."

A local family—the scientists' neighbors—appeared suddenly from across the road, adults and children running awkwardly in knee-high rubber boots, holding pots and pans over their heads as rocks continued to fall from the sky.

"They were screaming, and the father said, 'Please, take us with you,' " says Juan Duarte.

Duarte and Rafael Gonzales ran to the truck and started the motor. Salazar and Gil came out of the house and ran with the family of six. They all piled in on top of one another and raced down the dirt road away from the erupting volcano. Fernando Gil looked back at the mountain, watching wordlessly as they sped toward safety.

CHAPTER 2: BIRTH OF THE ANDES

IN COLOMBIA, three magnificent mountain ranges begin at the country's southern border with Ecuador and fan out to the north like the foot of a giant condor. The three sweeping cordilleras make up the Colombian Andes and are formed by the ongoing collision between two tectonic plates—pieces of the Earth's rocky crust that float on top of hot, ductile rock below the surface and move independently of one another. From the west, the heavy rocks on the Pacific Ocean seafloor creep deliberately east a few centimeters every year, while the South American continent, riding atop a plate of its own, stubbornly pushes in the opposite direction. For the last 200 million years, the two plates have been butting heads with enough force to create a three-tiered pileup that stretches over the entire western portion of Colombia and runs 5,000 miles along the curving South American continent.

The three mountain ranges have entirely different characters. In the west, the Cordillera Occidental follows the gentle crescent of

Colombia's Pacific coastline. The western mountains are a jumbled mess of basalt and ocean sediments that have been bulldozed into huge, irregular piles as the Nazca Plate, heavier and thinner than the continental crust, plows into and sinks below the South American Plate.

In the east, in the Cordillera Oriental, layers of billion-year-old sandstone, limestone, and shale bulge upward under the compressive force of opposing tectonic plates. The violent encounter causes the rocks to bend until they break along faults, and slabs of rock are thrust to the east over younger sediments, effectively reducing the width of the land.

Rising regally between the eastern and western cordilleras is the Cordillera Central, bordered by two great rivers that take deluges of tropical rain to the sea.

It is here that the mountains are alive.

Colombia's volcanic range is the result of the Nazca Plate, a 5-mile-thick slab of cold rock, sliding deep into the Earth's hot mantle. The plate does not sink straight down, but at an angle, where it ends up underneath the continental crust of South America. When the cold ocean slab reaches a certain depth, when the temperature and the pressure are just right, the ocean rocks release water into the surrounding mantle. Acting as a catalyst, water encourages the solid rock of the Earth's mantle to melt.

Over hundreds of thousands of years, the melting rock forms gigantic subterranean chambers of magma about 20 miles beneath the Earth's surface. And like an air bubble under water, the buoyant liquid begins to rise, pushing and fracturing its way through narrow conduits in the surrounding rock. Eventually, magma breaks through to the surface, and a volcano is born.

The volcanoes of the Colombian Andes are not unique. All over the world, when one tectonic plate sinks beneath another, towering, explosive volcanoes form on the surface above. In the Pacific Northwest of the United States, a small chunk of ocean crust called the Juan de Fuca Plate shoves its way into the mantle beneath the North American continent, giving rise to a chain of volcanoes called the

Cascades. Within the Cascades lie Mount Saint Helens, a 10,000-year-old cone that blew off 2,000 feet of its summit in 1980, and Mount Rainier, its million-year-old glacier-capped neighbor. In Japan, Indonesia, and Alaska's Aleutian Islands, the situation is the same, only the tectonic collision is between two ocean plates. One dense slab of cold ocean crust succumbs to thicker, younger rock and sinks into Earth, melting the rocks below and fueling the formation of volcanoes that explode through the ocean floor and create landmasses above the sea.

There are two main types of volcanoes: stratovolcanoes and shield volcanoes. *Shield volcanoes,* such as those found in Hawaii, consistently erupt liquid lava and form broad, sloping cones. *Stratovolcanoes,* like those of the Colombian Andes, are so named because of the way they grow; they are, literally, layered volcanos. A stratovolcano will experience many types of eruptions during its lifetime. In its youth, lava will pour from the stratovolcano's vent and radiate to form a sturdy base. As the volcano grows into adolescence, its magma becomes thick with mineral silica and has the consistency of fresh epoxy. The lava pours from the crater down the volcano's flanks, building a steep-sided cone. Over thousands of years, the thick, sticky magma solidifies and clogs the volcano's throat, and the young mountain's temperament begins to change. Beneath the cone, the pressure grows as magma continues to rise from great depths. During its ascent, the magma cools and releases gases, causing extreme pressure to build in the volcano's sealed throat. When the pressure becomes too much, the volcano explodes, pulverizing the cold rock in its vent and sending ash high into the atmosphere. Sometimes, a superhot avalanche of rock, ash, and steam races down the volcano's sides in what scientists call a *pyroclastic flow,* with speeds up to 300 miles per hour. With the volcano's vent once again opened by the blast, thick lava flows are able to pour more peacefully out of the crater decades, centuries, or millennia later. The process repeats itself again and again, and over tens of thousands of years, a tall gray cone appears, perfect in its symmetry: a giant mound of lava flows interlayered with crumbling pyroclastic deposits. Between large eruptions, the volcano may sleep peacefully, or it may be restless—coughing, sneezing, and occasionally clearing its throat by sending a puff of ash a mile high in an attempt to release pent-up pressure.

The volcano's evolution continues with alternating lava flows and pyroclastic flows until the stratovolcano reaches maturity; its flanks are now steep, its high summit tucked majestically into the clouds. But the volcano has arrived at this stature under a poor building code; the lava layers are strong and help secure the mountain, but in between the solid rock are layers of weakly cemented rubble. Over thousands of years, rainwater, made hot and acidic by volcanic steam, percolates through the porous pyroclastic deposits and turns them to crumbling clay. Gravity begins to cleave huge chunks from the mountain's steep slopes. After about a million years, the volcano's throat grows cold and solid, and damage caused by the constant wearing and weathering cannot be repaired with fresh lava flows. In the end, gravity and time will take their toll and erode the once-regal stratovolcano until it is level with the land.

As early as 20,000 years ago, human beings arrived in Colombia, traveling south along the Isthmus of Panama. By that time, the volcanoes of the Central Cordillera, once greatly explosive, had calmed down. Eruptions came as puffs of steam and small pyroclastic flows that likely earned the mountains respect but did not incite fear.

Between the equator and 10° north latitude, the land was warmed year-round by tropical sunshine. But unlike the steamy lowlands of Central America, the mountain ranges and high valleys in Colombia were cool and lush. In the Andes, the rain came twice a year, and twice a year the land would dry. In the lowlands, heat permeated the heavy air, while the very highest reaches of the cordilleras were permanently blanketed with snow.

The wide variety of climates offered homes to an amazing diversity of plants and animals. There were hundreds of species of mammals and reptiles, and thousands of different birds. Magnificent

Andean condors ruled the skies above the mountains, while colorful macaws, parrots, and toucans brightened the tropics. Colombia's flora was equally impressive: Hundreds of orchid species grew in the high valleys, and amazingly hardy plants, called *frailejones*, adapted to the harsh conditions above 12,000 feet.

Most of the people migrating south passed through Colombia and settled in what would become Ecuador, Peru, and Chile, where the great Inca and Nazca empires later flourished. Some, however, remained in Colombia, where the jagged mountains made it difficult for any one authority to control the area, and eight smaller factions settled in the high fertile valleys and plains that bordered the great cordilleras.

By the late 1400s, just before the Spanish conquest, the Muiscas were the most prominent indigenous tribe in Colombia. Estimated to have numbered 500,000, the agrarian and artisan clan covered a region more than 14,000 square miles along the Cordillera Oriental. At 8,000 feet above sea level, the land was lush and cool, the soil black and rich. The Muiscas enjoyed plentiful corn and potato harvests and possessed great wealth as the keepers of abundant salt and emerald mines dotting the foothills of the cordillera.

But it wasn't precious gems or rich soil that brought notoriety to the Muiscas. When the Spanish arrived on the Caribbean shore for the first time in 1499, the conquistadors were amazed at the wealth of the natives. Rumors abounded of a city of gold that lay deep in the land's interior. The emperor of the land, it was said, covered himself in gold dust, while his subjects threw precious gold statues and jewels into a ceremonial lake. The Spanish called the emperor *El Dorado*—the Gilded Man. Expeditions set out for the interior almost immediately, leaving from both the Pacific and Caribbean coasts and from Quito in the south. But the mountainous terrain was incredibly difficult to cross, and hundreds lost their lives to starvation, disease, and exhaustion.

By 1536, Gonzalo Jiménez de Quesada set out from the Caribbean coast on the expedition that would finally discover and conquer the Muiscas and the land where El Dorado was reported to exist. The Spaniard began his journey with 800 men, 50 horses, and 200 small boats that were to sail up the Rio Magdalena, the river that runs parallel to the western slopes of the Cordillera Oriental and feeds into the Caribbean Sea. The expedition met incredible horrors. Thin mountain air brought severe bouts of altitude sickness. The peaks were treacherous to climb, and the swamps were full of disease-carrying insects. Food was scarce, and the Spaniards found no Indians to steal from, bringing many to starvation. In the end, only 200 of Jiménez de Quesada's men made their way to the eastern foot of the Cordillera Oriental—and into the territory of the Muisca.

Even with three-quarters of his men lost, Jiménez de Quesada had little trouble subduing the peaceful Muiscas, who at the time were divided into two clans: the Zaque in the north and the Zipa in the south. In 1598, Jiménez de Quesada established the city of Santa Fe de Bogotá on top of Bacatá, the principal city of the Zipas. But the bounty of gold artifacts that the Spaniards pillaged from the natives failed to satisfy Jiménez de Quesada, who was certain that there must be more gold buried in a secret hiding place. The conquistadors had planned to capture and torture the Zipa leader into confessing where the treasure was hidden, but he was killed in battle. His successor also died without divulging the source of all the gold. The Muisca towns and their treasures fell to the conquistadors, but the golden cities and rich mines were never discovered.

The reason for this was simple. The Muisca gold had been purchased from distant tribes with salt, emeralds, and fine cotton fabrics. The mountains in the west—hundreds of miles and two mountain ranges away—were rich with gold deposits that had formed from the heat of the colliding Nazca and South American tectonic plates. The gold had concentrated in hot fluids that circulated through fractures in the rocks as the mountains formed. Over tens of thousands of years, tropical rain eroded the mountains, exposing bountiful veins of gold and sending weathered rocks and minerals into streams and rivers below, where the heavy precious metal became ripe for harvesting by the native tribes.

As conquerors of the land east of the Cordillera Oriental, Jiménez de Quesada's men were nowhere near the source of the Muisca gold. But the ceremony that gave rise to the legend of El Dorado *did* come from the Muiscas and had been a tradition for hundreds of years before the Spanish arrived. Just 30 miles north of the newly crowned capital of Santa Fe de Bogotá lay a small, perfectly round lake, carved into the cordillera foothills. This lake would become an obsession to the Spaniards for the next four centuries.

Laguna de Guatavita was a lake full of gold.

For hundreds of years before the Spanish had arrived, the 100-foot-deep, tree-lined lake had been the center of great ceremony. When a new Muisca chief was inaugurated, his naked body was first painted with a sticky substance and then covered entirely with gold dust. The new ruler would descend the lake's steep slopes and take to the still water aboard a small raft. The raft was adorned with fantastic gold artifacts and jewels, and all along the narrow lakeshore, the ruler's subjects would toss gold and gems into the lake in a ceremonial offering. At the ceremony's end, the new Muisca ruler would plunge into the lake's cold, emerald green water to gather the power of the gods.

Soon after the Spaniards invaded the Muisca land, they made their first attempt to drain Laguna de Guatavita. Human chains bailed the lake with hollowed gourds, but failed to lower the water level enough to expose more than a few feet of shore. Decades later, a giant notch was carved in the lake's rim that collapsed around the crew during construction, killing many of the workers. Each attempt produced just enough gold to entice further generations to try and capitalize on the lake's riches, and in 1898, the Company for the Exploitation of the Lagoon of Guatavita set out to empty the lake by carving a tunnel into the lake's center. Portals would regulate the flow, and mercury screens were set inside the tunnel to capture emeralds and gold. The scheme seemed to work. The tunnel hit the lake center, and water flowed out, emptying the 100-foot-deep lake. Immediately afterward, the lake bed was knee-deep in slimy mud and impossible to walk across. So the company waited. By the next day, the clay-rich mud had turned to cement in the hot sun and was impossible to penetrate. The mud clogged the portals and sealed the tunnel, and Guatavita soon filled with water to its original height. A few hundred British pounds of gold collected in the first day were auctioned, but the company went bankrupt soon after. In the following decades, several more attempts were made to drain the lake with mechanical earth-moving machines, until finally, in 1965, Laguna de Guatavita came under the protection of the Colombian government.

By then, the 400-year search for the legendary city of El Dorado had taken a ruthless toll on the indigenous population of Colombia. The Spanish did not give up their greedy search after the Muiscas failed to provide the boundless riches they were seeking. The conquistadors and their successors eventually eradicated nearly all the indigenous peoples of the northern Andes. Today Indians account for only 1 percent of the entire Colombian population.

The land of the Colombian Andes was quite different from the steady European soil from which the Spaniards had come. Besides being littered with gold and strewn with volcanoes, the ground itself was alive—and it was deadly. The violent earthquakes, or *terremotos,* came without warning, caused by the same tectonic forces that had built the volcanoes. But unlike the small eruptions of Colombian volcanoes, the earthquakes caused hundreds to die, crushed under the homes and churches they had built to protect themselves. In an instant, the ground would rattle and shake and then begin to sway and rock as if riding on a ship at sea. Brick and stone buildings crumbled to the ground around their inhabitants, cracking and collapsing into giant tombs.

The first earthquake recorded by the Spaniards took place on September 1, 1530, at 10 o'clock in the morning. The massive tremor shook the Caribbean coast from Cartagena to Caracas, crushing newly erected Spanish villages when sandy soil turned to liquid under the vibrating seismic waves. Over the next 400 years,

300 significant earthquakes rocked the northern reaches of South America. In the capital of Bogotá, churches tumbled and shanty-towns fell to pieces. In Popayán and the sugar-producing lands near Cali, hundreds were killed by falling structures. The prosperous cities of Medellín and Manizales to the north were leveled and rebuilt over and over again. Lethal tremors tormented the city of Pasto at Colombia's southern border with Ecuador. As the ruined villages mended and rebuilt and the populations grew, more and more casualties occurred with each quake, and it was often impossible for help to arrive because of the mountainous countryside. Although the land rumbled endlessly, there were no scientists in the entire country of Colombia, and the people blamed God's wrath for the perpetual destruction.

The turmoil beneath the earth's crust seemed to rise to the surface and infect the land's human inhabitants. Gran Colombia, a territory that comprised what is now Colombia and Ecuador, severed ties with Spain in 1819 with the help of Simón Bolívar. But with Bolívar gone off to conquer other parts of South America, the vast territory soon split in two, and the Liberator's hopes of a sacred union encompassing all of the land he had freed were shattered. The nineteenth century proved to be exceptionally bloody for Colombia, culminating with the devastating War of a Thousand Days in 1899, which left more than 100,000 dead. The wars were fueled by economic stagnation that resulted from the incredible difficulty of getting from one place to another in the young country. It was not a matter of great distances between economic centers—Bogotá and Medellín were only 120 miles apart, and the port city of Cartagena was less than 400 miles from the capital. But the jagged mountains and valleys served to isolate the population centers, and the cost of shipping goods between cities was extremely expensive because of a lack of roads. Man and mule provided the only mode of transporting cargo through the cordilleras. The range of climate zones in each region enabled the departments or states to be agriculturally self-sufficient, and they grew as individual entities with strong feelings of separatism and would remain as such through the millennium.

By the early twentieth century, civil wars had subsided and

Colombia was experiencing a period of quiescence. Times were especially good in Antioquia, the department that comprised the northern regions of the central and western cordilleras. The area contained 70 percent of the country's gold, and its capital, Medellín, was a thriving industrial center.

In 1910, not far from Medellín, in the village of Yolombó, a young boy named Jesús Emilio Ramírez walked the rolling hills of his family's *finca*. His mind would wander, and he would gaze up at the tall mountains before him. Jesús Emilio loved the earth. He was amazed by the way the mountains jutted powerfully up from the valley corridors. He would delight when the *terremotos* would come; the wooden floor would rattle beneath his feet, and his entire house would sway on a seemingly liquid earth. And although he had not yet seen them, he fantasized about the active volcanoes of the Central Cordillera that were said to breathe fire and explode with fury. To Ramírez, the earth was alive, and he longed to understand why it was so.

In 1922, when Jesús Emilio Ramírez was a teenager, the first seismograph was delivered from Spain to the Jesuit university in Bogotá. The instrument was placed in the Jesuit high school in downtown Bogotá, where the priests took daily readings of changes in movements of the atmosphere, hoping to predict the onset of severe storms.

The weather seismograph was interesting, but it frustrated Ramírez. With the exception of a few priests studying meteorology, there wasn't anyone in Colombia working in the physical sciences. He decided to join the Jesuits, a Catholic order of priests that was world renowned for its scientific pursuits. He spent four years studying humanities in Bogotá, then left for the United States and the University of Saint Louis, where he earned a doctorate in seismology in 1939.

Armed with the latest scientific understanding about the dangerous *terremotos*, Ramírez returned to Colombia and, in 1946, founded the Los Andes Geophysical Institute on the back lot of the Jesuit university in Bogotá. The institute, tucked into a small four-room building, was dedicated to earthquake research, but Ramírez encouraged students to tackle all types of earth sciences. That same year, a group of Jesuits in Germany gave Ramírez the first earthquake seismometer to be placed on Colombian soil. He painstakingly assembled the device, taking great care with its brass springs and wires and fitting it gently into its polished wooden case. The seismometer needed to be installed in a secure place, free from the vibrations of a bustling city, and Ramírez chose the foothills above the Jesuit elementary school. The tall priest acted as foreman while six workers dynamited a deep tunnel into the bedrock. Afterward, still wearing his long black robe, he joined the men as they began to clear the way into the mountainside.

The workmen built two rooms into the excavation. In the back room, the heavy German seismometer was set into the ground. The seismograph, with its needle poised to record what the seismometer felt, was set on an oak desk in the front room. Ramírez instructed the workers to build a facade over the front of the tunnel on which they painted a map of North America to the left of the door and one of South America on the right. A 6-foot bronze globe stood in front of the tunnel entrance, and in brass letters above the heavy wooden doorway a proud sign spelled out Ramírez's dream: Instituto Geofísico de los Andes Colombiano.

Ramírez personally went to the mountainside each morning to collect the seismograph data. He spent hours teaching anyone who was interested what the earthquake signals meant. Spreading out the long black sheets, Ramírez explained that he could tell if the earthquake scribbled into the carbon-coated paper was small or large, and whether it was from just below Bogotá or as far away as China. It was similar to the waves that are produced by throwing a pebble into a still pond, he would explain. The waves radiate from the center and take longer to reach the pond's edges. If the seismic waves drew very tight scribbles, the epicenter of the earthquake was close by. But if the recorded signal appeared as wavy lines that were far apart, the seismometer was reacting to the slow heaving of the earth in response to a faraway quake—something a human would never feel, but the sensitive seismometer could. The size of an earth-quake could also be estimated by the seismograph reading. If the earthquake was very strong, the scribbles would extend several vertical inches. And if the tremor was small, the signal may rise just a bit above the horizontal lines.

In the 1960s, nuclear proliferation became a turning point for the study of seismology. When a nuclear bomb was detonated, it sent shock waves through the surrounding rocks that were picked up on seismographs just like earthquakes. Soon, the entire world was being wired with seismometers to keep a subterranean eye on nuclear blasting. In Colombia, Ramírez wasted no time making sure his country would benefit from the allied International Seismic Network. The superpowers may have been worried about an arms race, but to the padre, it was a scientific jackpot. Ramírez ended up with six seismic stations—compliments of the allies—that were placed around Colombia in the most earthquake-prone areas. While governments monitored for nukes, the Jesuit scientists collected precious data on Colombia's earthquakes. The institute was in charge of all seismic research in the country, and for the next two decades, Ramírez vigilantly pushed to keep his country at the forefront of earthquake science. Unfortunately, violence by guerrillas and emerging drug cartels plagued the country and kept foreign researchers from conducting scientific studies and installing state-of-the-art equipment in Colombia, and new research was mostly limited to the meager funding available from the Jesuit university.

The study of earthquakes was not Ramírez's only passion. Ever since boyhood, he had delighted in the legend of the fire mountains told by the Tolima Indians who had inhabited the central cordillera:

> *In white earth where the gods live, there are seven peaks with great hollows in their centers that penetrate to the center of the world and their white pinnacles reach the heavens. It is in this celestial land were the gods, wizards, and oracles, surrounded by beautiful maidens, meet and discuss the future, predictions for the weather, and coming misfortune.*

It was in this sanctuary that one particular wizard—a drunk and evil one—would violate the most beautiful of maidens, claiming in his favor a divine right. One day, he chose as the object of his desire a young maiden who belonged to the caste of Queen Dulima. The queen would not let the horrible wizard take her lovely maiden, and she and the other maidens devised a plan.

They served the wizard intoxicating drinks made of fermented flowers. The young maiden then enticed the wizard into a cave, fooling him that she would give in to his depravity. Her maiden friends had privately prepared large rocks and bits of large crags, and at just the right moment, the beautiful maiden escaped and her friends covered the cave's entrance with heaps of earth and rubble, silencing the wizard.

Or so they thought.

The evil wizard was buried alive, and his infernal magic roared and trembled under the feet of the terrified inhabitants. Then sprang forth the entrails of the mountain in columns of black smoke, rivers of sulfur, and tongues of fire. And then came the shaking of the cavern, welling up out of three tumultuous fountains of water: one bitter, another boiling, and the third cold like thawed snow, springing forth to the surface of Volcán Machín.

The story of Volcán Machín, and the seven white peaks, which included both Nevado del Ruiz and Tolima, was a favorite of Ramírez, but he longed to know more. Over the years, he had painstakingly gathered eyewitness accounts of Colombia's volcanic activity dating back to the Spanish conquest. There were hundreds of detailed records by priests, monks, historians, naturalists, and soldiers. Of Colombia's thirty volcanoes, seven had experienced eruptions since the mid-1500s. In 1816, near Popayán, Volcán Puracé blew ash and andesite shrapnel. In 1899, Doña Juana creaked and rumbled and colored its rivers yellow with sulfur. Cumbal and Tolima sporadically erupted ash and steam, and in 1580, at the southern border of Colombia, Volcán Galeras rained ash onto the city of Pasto and continued to undergo periods of unrest each century thereafter.

For the most part, the eruptions of Colombian volcanoes created excitement and wonder but caused little damage. There was, however, one exception. The giant Nevado del Ruiz had spread terror and destruction twice in recorded history. In both cases, after relatively small eruptions, the volcano's glacier cap quickly melted and sent torrents of ice, rock, and mud down steep valleys. Completely without warning, the low-lying cities at the base of the mountain were buried by millions of tons of debris-laden mud. The number of people killed was impossible to know, as entire villages were swallowed by the moving earth.

Ramírez also discovered that many of the early observers had made intuitive connections between earthquakes and volcanic eruptions. It was hard for Ramírez to see how it could be possible. Did rocks fracturing along deep fault lines create heat to fuel volcanoes, or were the burgeoning volcanoes violently rupturing Earth's crust? In the early 1960s, the scientific community—in the most amazing geological discovery of the century—would begin to uncover the alliance between the two incredible phenomena.

The idea had been under debate since 1912, when Alfred Wegener, a German geologist, meteorologist, and explorer, published a theory he called *continental drift*. Drifting landmasses were Wegener's explanation for the puzzle-piece fit of Africa and South America, the matching 200-million-year-old reptile fossils on each continent, and the way that mountain ranges and volcanoes buckle up when continents steamroll over rigid ocean crust.

Wegener's theory seemed outlandish at the time, and he was ridiculed in the scientific community and blasted by the church. The German had failed to convince any of his detractors because he could not describe a mechanism powerful enough to move giant continents. Wegener's theory sat on the shelf of oddball ideas until 1962, when scientists discovered that new seafloor was being created in the middle of the Atlantic Ocean. From a great fissure that extended from the Arctic Circle to the continent of Antarctica, thick basalt lava poured forth, paving the bottom of the sea. The "spreading center," as it was named, was effectively pushing the Americas away from Europe and Africa.

While new seafloor was being created, old, colder seafloor in the Pacific was sinking back into the earth, and along the coast of South America, unmanned submarines found trenches 30,000 feet deep. Wegener had been partly right. But it wasn't just the Earth's continental masses that were moving, it was the planet's entire outer shell. The fuel that drove the giant plates was the Earth's fiery core, a blistering 7,000-degree mass of solid iron and nickel that was still holding the heat from the Earth's formation. The outer shell, exposed to the cold temperatures of space, cooled first and formed a hard crust. But the heat still trapped inside wanted out, and it broke its way through the ridged rock, spilling out in oozing flows of hot lava along the seafloor and fracturing the Earth's crust into giant, irregularly sized plates.

To Jesús Emilio Ramírez, the theory was a beautiful symphony that could only have been conducted by the hand of God. Like the divine blending of orchestral voices, magnificent mountains rose toward the heavens. Living volcanoes exploded in tremendous crescendos, and in a crash of great cymbals, scorching Earth broke from its shell in a thunderous hymn of renewal.

In January 1981, the 75-year-old priest, who had devoted his entire life to helping Colombians understand their volatile country, died at his home in Bogotá—just five years before Nevado del Ruiz would erupt in one of the deadliest volcanic disasters in history.

CHAPTER 3: GHOSTS IN THE KITCHEN

NEVADO DEL RUIZ, COLOMBIA:
DECEMBER 16, 1984

THE COLD DECEMBER WIND sounded like it would blow right through the wooden mountaintop *refugio*, or refuge. Inside, Maria Elena Vivas warmed her mittened hands over the gas stove in the kitchen of the rustic hotel. It was 10 P.M., December 16, 1984—nearly a year before Fernando Gil began his job with Bernardo Salazar—and she was the only one still awake in the shelter, high on the western slope of Nevado del Ruiz. The rustic, five-room hotel was operated by the National Park Service and run by a stocky, bearded *Andinista*—a mountaineer—named John Jairo Gutierrez.

Vivas had been working at the hotel for a year, since December 1983. She cleaned the guest rooms and bathrooms and scrubbed pots and pans and dirty dishes in the kitchen after meals. She swept continually, but she never seemed to be able to free the wooden floors from the dust tracked in by visitors. During the day, she loved being surrounded by guests. But at night, when everyone was in bed, the shelter would creak and moan, and Vivas would long for her parents' home in Manizales and for a much warmer place to sleep. "It was always very, very cold," Vivas says.

The *refugio* catered to mountain climbers who came to conquer Nevado del Ruiz's steep, icy slopes. Some would even hike up the snow-topped glaciers with their skis and slalom down. It was mid-December. The rainy season in the Cordillera Central had just ended, and there was a fair amount of fresh snow. But with the holidays so close, the hotel had few visitors.

The *refugio* was not for the luxury traveler. The two-story structure rested at 15,000 feet, just 2,000 feet below the summit of the volcano. Although there were propane tanks for cooking and the building was wired for electricity from a nearby dam, "the power went out all the time, and the generator that was supposed to back up the power was always broken," Vivas says. Water came from a nearby well through pipes that were frozen most of the summer because of gale-force winds that blew freezing air from the east. When the water did come, it stank of sulfur and could only be used for cleaning and some cooking. Drinking water had to be brought up in 10-gallon plastic jugs. Visitors would often become nauseated and leave the shelter with a throbbing headache soon after arriving.

The guest rooms in the two-story building were sparsely furnished with hard mattresses, and the walls were mostly bare. There were electric heaters, but warming the air thinned the oxygen even more at such a high elevation, so Gutierrez, called "J.J." by his friends and fellow climbers, seldom turned on any heat. The hotel did not take reservations. The place would fill to capacity on the weekends, but during the week, Gutierrez rarely had to turn anyone away.

The Nevado del Ruiz volcano was the crowning jewel in the Parque Nacional Los Nevados, a 225-square-mile collection of five volcanoes and surrounding mountains that made up the highest part of the Cordillera Central. On a clear day, the view from the *refugio* was spectacular. In the front yard, the dirty tip of a 100-foot-deep glacier tongue reached from the summit of Nevado del

Ruiz. The volcano, rising to the east, was flat and broad on top and covered by 9 square miles of ice that crept down eight radiating valleys like the tentacles of a giant octopus.

In the evenings, the hotel's high-ceilinged dining room would fill with exhausted mountaineers and skiers who drank cup after cup of *tinto*—opaque black coffee loaded with sugar. After *tinto,* there were hefty shots of brandy, rum, and *aguardiente,* a Colombian anise liquor.

That night, Maria Elena Vivas walked into the empty dining room and began clearing dishes from the long wooden table in the center of the room. Suddenly, the floor began to vibrate beneath her feet with an almost imperceptible shaking that quickly intensified.

Vivas grabbed the corners of the table and tried to hold it steady, but her arms quivered from the vibration. "The entire *refugio* was rattling," she says. For another thirty seconds, she watched, horrified, as cups and dishes inched toward the edge of the table. Then they stopped.

Slowly, she released her grip. The dim light fixture that hung from the ceiling swayed in a foot-long arc, casting eerie shadows on the dark wooden walls. Vivas felt as if her heart would burst from her chest. She sat down perfectly still on one of the wooden chairs and listened for noises, but the house was quiet except for the wind. No one in the *refugio* had awakened.

Vivas slept fitfully that night. She rose before dawn as usual and made her way to the kitchen, carrying an oil lamp and stepping timidly down the narrow hallway. J.J. Gutierrez was already in the kitchen and had started to brew some coffee.

"I was scared, and I told J.J. the house was rumbling like there were ghosts walking right through the kitchen. At first, he didn't believe me," says Vivas.

From the kitchen, Vivas heard the guests coming into the dining room, so she started to make breakfast. It was still completely dark, not yet 5 A.M., and she was grateful for the chattering voices in the dining room. As she held a kettle to the waterspout on the giant plastic jug that sat on the counter, the floor began to vibrate. Vivas dropped the kettle and ran out into the dining room. The mountain climbers were on their feet. The entire hotel was vibrating, and the noise of everything rattling created a low-pitched hum.

Vivas remembers Gutierrez running in the front door, his face white. "He knew a lot about Nevado del Ruiz from the campesinos whose families had lived there for hundreds of years," Vivas says. "They told him the volcano hadn't always been sleeping."

Gutierrez and Vivas were frightened. The volcano was rumbling. The quiet, beautiful volcano could be waking up, and here they were, just a half-mile below the volcano's summit and the gurgling throat of Nevado del Ruiz.

For the next several days, the low, humming vibrations shook the *refugio*, and each day the tremors seemed to come more and more often. The shaking was not violent and did little more than rattle a few cups on the kitchen shelves. Gutierrez and Vivas gradually became accustomed to the rumbling. "We told the people who came and were frightened that that's just how it is when you're on a volcano," says Vivas.

Three days after Vivas had felt the first tremor, two mountain climbers went to the summit and reported back to Gutierrez that the crater of the volcano, recently covered by ice, was now clearly visible. There was a strong smell of sulfur, they said, and the snow on the top was covered with tiny yellow crystals.

Gutierrez decided that he needed to see what was happening for himself. The next day, he left the hotel at 4 o'clock in the morning with Luis Fernando Toro (known only by his nickname, "Bis"), the most famous and knowledgeable *Andinista* in Colombia. It took the seasoned climbers just two hours to make the mile-long, 2,000-foot-high climb over ice and rock. When they reached the top, they were stunned by the incredible change in the once-smooth summit. The crater was a deep, 500-foot-wide muddy pool. From several distinct holes in the mud, thin columns of wispy steam billowed into the atmosphere. The air stank of sulfur and burned their nostrils and eyes. Gutierrez felt like he would suffocate from the acid in his throat and repeatedly spit on the snow, trying to get rid of the burning sensation in his mouth.

Gutierrez went back to the *refugio* and radioed the park authorities a report of what he had seen. The park authorities called the Civil Defense for the state of Caldas, and soon word began to spread that the beautiful and tranquil Nevado del Ruiz was awakening.

One of the first people to hear of the volcano's restlessness was Pablo Medina, the 45-year-old vice president of the financial corporation of Caldas. Medina, a handsome, distinguished man from one of the oldest families in Manizales, was a coffee grower himself and very interested in protecting the wealthy region from harm. He was one of the first people to reach the summit of the volcano in 1966, and he had climbed up and skied down the mountain's steep slopes dozens of times since. It was worrisome that the volcano was growing active, but not something that Pablo Medina was afraid to face head-on.

It was a painful and frightening time for Medina's beloved country. Abundant with natural beauty, gold, emeralds, petroleum, and an endless expanse of rich black soil, Colombia seemed to be in a constant state of self-destruction. Struggles between Colombia's two ruling parties, the Liberals and Conservatives, had fueled many civil wars in the twentieth century. The worst, *La Violencia,* a five-year war that ended in 1953, left more than 200,000 dead.

Because of continuing disparities between the small ruling elite and the peasants, guerrilla movements that began in the late 1940s were raging out of control by the 1980s. Four major factions, with ties to Cuba and the Soviet Union, terrorized the country, kidnapping the rich for ransom, murdering peasants in the countryside, hijacking buses and airplanes, and bombing electrical towers and oil platforms.

And yet another poison was spreading throughout the country. In the 1970s, Colombia had become a key stop in the trade of Peruvian and Bolivian cocaine. Thousands of refugees fleeing guerrilla violence in the countryside quickly became involved with the drug cartels. The drug lords, who had earned billions through their illegal trade with the United States, became wealthier than the traditional elite, and to ensure that they were never incarcerated for their crimes, they paid off judges who would take bribes and murdered those who wouldn't.

By the 1980s, guerrilla rebels and drug traffickers were infiltrating Colombia's rain forests and waging war on each other. The guerrillas started to tax the cocaine growers, while paramilitary groups, hired by cattlemen who had joined forces with the drug traffickers, began to execute the rebels.

But throughout the seemingly endless violence, the Cafetera region had remained calm, in large measure because of the tireless work of Pablo Medina and other concerned leaders of the Cafetera. They made sure that the laborers in the coffee fields were paid decent wages and that their families were well cared for by the wealthy landowners. There was little unrest or poverty in the region to fuel sympathies toward the guerrillas or desires to align with the drug lords.

As with the other challenges he had faced, Medina took a proactive stance toward Nevado del Ruiz. He convinced the financial corporation of the coffee growers to lend him the empty eleventh floor of the Banco Cafetero building in Manizales. And knowing that Colombia had no native volcano experts, he put out a call to the United Nations. He told them that the regions surrounding the Parque Nacional Los Nevados had an active volcano on their hands. Then he readied for the hordes of international experts who were sure to come.

Meanwhile, local and regional scientists began to trickle in to *Piso Once*. Pablo Medina's biggest score came from the Central Hydroelectric Company of Caldas, or CHEC. The corporation had several dams on rivers near the volcano that produced power for the state of Caldas, and they were worried that their power plants could be damaged by an eruption. They also had been doing geothermal research in the area around the volcano and employed the closest thing to a volcanologist that the entire country of Colombia had to offer: a young, vibrant geologist named Marta Lucía Calvache.

Calvache returned to Manizales in the first week of February 1985, after spending three months in New Zealand. She walked into the CHEC office in Manizales early Monday morning, and was immediately met by her colleague Nestor García, a chemical engineer. "As soon as Nestor saw me, he didn't even say, 'Hi, Marta, how are you?' He said, 'Nevado del Ruiz is active, we have to go to Nevado del Ruiz,' " Calvache says.

Calvache had never seen García like this. Normally subdued, he was shaking with excitement. García told her that he had gotten a call from Bernardo Salazar, a civil engineer they both worked with at CHEC. The bosses at CHEC were turning over the problem of the smoldering volcano to a team of three including Salazar, Calvache, and García.

"To the bosses at CHEC, it just seemed like the normal thing to do, to put us to work on the volcano," says Calvache. But still Calvache couldn't help but wonder why.

García told her what all of the scientists in Colombia already knew: Marta Calvache had more experience with volcanoes than anyone in the country. To Calvache, the idea seemed preposterous. She was barely getting started in her career, and she had never had any training on an active volcano. The three months she'd spent in New Zealand taught her about the geology and chemistry of the country's geothermal energy fields. But dealing with the potential hazard of an active volcano was completely beyond her expertise.

Her mind was reeling as Nestor García began to explain what had happened since the people at the *refugio* first felt the mountain rumbling. Slowly, uneasily, Calvache began to accept that, at 26 years old, she was considered Colombia's leading expert on volcanoes.

In 1976, Colombia was enjoying a hefty oil boom, and 18-year-old Marta Calvache had several relatives who worked as geologists for Ecopetrol, the national oil company of Colombia. She was just about to start college, and since the job market was looking good for careers in the oil business, Calvache enrolled for classes as a freshman geology major at the National University of Bogotá. Upon her arrival, Calvache instantly made friends with Maria Luisa Monsalve, another geology major. Never before had the program had more than one woman, and this year there were eight.

Calvache and Monsalve took classes together, making sure to pick electives that would help them in their careers with Ecopetrol. They studied the plentiful oil reserves in Colombia east of the Cordillera Oriental, where thick layers of 100-million-year-old sedimentary rocks trap oil and gas reserves in porous sandstone and limestone. Calvache and Monsalve learned to identify places that might be rich in oil: the folds and faults in ancient rock beds that trap and concentrate the valuable crude.

The two women worked and studied for three years, when suddenly, after having problems with some National University graduates, Ecopetrol announced it would no longer hire anyone from the university. Calvache and Monsalve were crushed. "All the classes we took were meant to prepare us to work with Ecopetrol—the *only* oil company in Colombia," Calvache says. So the women sat down and tried to figure out what to do next.

A month later, Monsalve's brother told them that CHEC was looking for students to work with their new geothermal program. The job would take them to Manizales, and they would be much closer to Monsalve's family in Armenia. "For Maria Luisa, it was the perfect solution. Her mother had just died, and she wanted to be near her father and brothers. But the thing was, we didn't know anything about volcanoes or geothermal energy," says Calvache. But since jobs anywhere in their field were hard to come by, Calvache soon agreed. "I was not very convinced, but it was the best situation for Maria Luisa," she says.

The women moved into small apartments in Manizales, a bustling city surrounded by green mountains covered with coffee plantations. They would be working on a project with a group of Italians, exploring for geothermal energy, trying to find ways to tap into the volcanic heat and turn it into steam to run turbines that would generate electricity. Every day, they hiked for miles and miles around the lush, jagged Andes that surrounded the massive volcano and trekked across sheer walls 500 feet high formed by million-year-old lava flows. The two best friends spent their days breathing the cool Andean air and learning everything they could from the Italian scientists. Searching for geothermal energy, they found, was like searching for gold, oil, or precious gems: you looked for certain types of rock formations. In the case of geothermal fluids, the rocks to look for were volcanic, porous, and fractured by faults.

In the fall, Calvache and Monsalve went back to Bogotá. They finished their classes, graduated, and returned to Manizales to start full-time jobs at CHEC. For the next three years the two women worked together until Monsalve was sent to Paris to study geology, and Calvache was sent to New Zealand to learn the geologic workings of geothermal energy fields. Now, just five years out of school, Calvache had been given a staggering responsibility.

She left the office at 3 in the afternoon and went home to pack the things she would bring to the volcano. "Nestor told me to dress warmly, get some sleep, and to make sure to eat something before we left." She shoved two thick wool sweaters into her backpack. She set out a pair of jeans, some thick socks, and the thin street boots she always wore in the field. In order to reach Nevado del Ruiz's summit before the sun turned the snow to slush, making it impossible to walk across, they had to leave at 1 A.M. the next morning. At 5 o'clock in the evening, with the sun still shining in her bedroom window, she tried to sleep.

It was impossible. She got up and went back to bed several times but had no luck. By 10:30 P.M., knowing she had to get up in two hours, her mind was reeling. What could she do about an active volcano? What would she do at the summit? "I knew that it's not so easy to climb to five thousand meters," she says. Would she even be able to make it to the top?

She lay in bed with her eyes wide open and stared at the dark ceiling. She tried to eat, but couldn't, tried to sleep, but couldn't. Calvache had never run from a challenge, and she didn't plan to now, but she had never faced one quite this big before. Ten minutes after Calvache had finally fallen asleep, Nestor García was at her apartment door.

The ride into the national park was a dark one, along endless

winding roads. Bis, the *Andinista* who would be their guide to the
summit of Nevado del Ruiz, drove the truck hurriedly up the mountain. In the front sat Pablo Medina, Bis's old climbing buddy. Bis
seemed to talk nonstop, and only about one thing: Nevado del Ruiz.
Calvache and Nestor García sat quietly in the back.

As Bis careened up the mountain, the air became colder and
colder and a bright half-moon offered a view of the road ahead. It
was a giant zigzag that went up the steep, barren approach to the
refugio. Calvache stared out the window and surveyed the bleak
landscape; lush green vegetation no longer covered the mountains—
just gray, desolate rock.

They arrived at the hotel and piled out of the truck. It was
4 A.M., and the windows of the brown two-story building were
dark. The wind howled across the barren landscape. Bis knocked
loudly on the front door, and a sleepy J.J. Gutierrez appeared a
minute later.

The two *Andinistas*, Bis and Medina, sat in wooden chairs surrounded by a pile of equipment and laced their heavy climbing
boots. Bis wore heavy black pants and a puffy red jacket. When he
finished lacing his boots, he wrapped blue nylon sleeves around the
bottom of his pants and tied them under the soles of his boots.
Sunglasses dangled on a thin cord around his neck. Pablo Medina
was similarly decked out.

Calvache and García got ready in less than five minutes. "We
didn't have the right clothes or the boots—nothing. And of course
nothing like an ice axe."

Bis surveyed the two scientists and let out a big laugh. "Okay,
guys, let's see what kind of *Andinistas* you are!"

The four of them stepped outside and began to hike up the trail
that would take them to the top of the volcano. It was 5 A.M. and
still very dark. Bis shone a flashlight on the ground, but it only illuminated a narrow path, and Calvache had to stare hard at the
ground to make sure she didn't stumble over rocks protruding from
the crusty snow. She and García followed in the stair-step footprints
left by their guides' heavy mountaineering boots.

Once they got above the rocks and were on the glacier, Bis and Medina stopped. They were 100 yards ahead of Calvache and García, who were having a difficult time as their flat-soled boots slipped on the ice. The sun rose on the faraway horizon and turned the glacier a dull pink.

When the two young scientists reached the mountaineers, they were breathing hard. They had stayed fairly warm during the vigorous hike, but their shoes were wet and their hands and feet were freezing. Bis regretted that he didn't have an ice axe for either of them, an indispensable tool for hiking on glacier-covered terrain. They didn't have harnesses or enough gear to rope the four of them together. But Bis knew that it was important to get these two scientists to the top of the volcano, so he gave the novice climbers a ten-minute crash course in alpine mountaineering.

He was happily surprised. The two were tough and in excellent shape. Within three hours, the party reached 17,000 feet. It was the first time García and Calvache had ever seen the top of the volcano. The sun was climbing and it illuminated the glacier-covered peak, which now sparkled with yellow sulfur crystals. The broad summit of Nevado del Ruiz spanned a half-mile. Toward its northern edge lay the steaming mouth of the restless volcano.

Calvache and García stood still, inhaling the thin, sulfur-laden air and staring at the living mountain. A dozen fountains of white steam escaped from its massive crater, hissing softly in the wind. It was true—Nevado del Ruiz was no longer sleeping.

Calvache knew what was causing the fountains of steam, called *fumarolas:* they were similar to the geysers that she had studied in New Zealand. In other parts of the crater, mounds of glistening mud stood several feet high like giant anthills, with small center craters where warm silt bubbled to the surface.

García hiked down into the crater toward the billowing columns of steam. "I remember Nestor took a regular thermometer from the lab and just tried to stick it in a fumarole." But the gas was so hot that García couldn't get close enough to get a reading on the small thermometer. He tried several times and finally gave up.

Calvache walked to a flat spot in the crater. She took out her yellow field notebook and began to sketch the crater and the positions of the fumaroles. She noted the smell, the color, and the height that they reached before they dissipated into the thin atmosphere.

Bis and Pablo Medina told Calvache that they had been to the crater dozens of times over the last decade, and always, an endless expanse of ice had covered the summit, with only a shallow depression marking the mouth of the volcano. Now, the steaming crater was as wide as two soccer fields and as much as 100 feet deep.

The sun rose higher and turned the patchwork of small farms surrounding the volcano a brilliant emerald green. The sky was a pale blue streaked by soft wispy clouds.

Calvache stood on the glacier and looked to the south. At the far reaches of the summit, La Olleta, the remnants of an older, dormant crater of Nevado del Ruiz, stuck up like a giant snow-covered tree stump. Behind La Olleta sat two more volcanoes: Cisne and Santa Isabel. And farther in the distance emerged the perfect snowcapped cone of Tolima. Calvache slowly turned clockwise. To the west, she surveyed the Cafetera region and the wealthy city of Manizales, rising proudly on a steep ridge surrounded by endless coffee plantations. She turned to the north and saw the sleeping volcano Cerro Bravo with its smooth, sloping flanks. As she turned east to face the rising sun, Calvache's eyes followed the Rio Lagunillas as it cut a deep valley through the seemingly endless expanse of mountains. Calvache knew that the river ran all the way to the base of the mountains, across the flat Armero plain, and entered the giant, winding Rio Magdalena that flowed north to the Caribbean Sea.

Looking across the horizon at what seemed like the entire world below, she forgot how cold she was. Her mouth and throat burned with acid, she could barely breathe, and she hadn't slept for more than ten minutes the night before, but none of that mattered.

This is my new job? she thought, squinting in the sunlight reflected off the glacier. *How could I possibly be so lucky?*

*   *   *

While Nestor García and Marta Calvache continued to visit the crater each week and look for changes in Nevado del Ruiz, Pablo Medina tirelessly solicited more support for their efforts. Soon he had corralled the chamber of commerce and several other corporations in Manizales. He was able to interest a meteorology professor from the National University of Manizales and several other local academics.

On his own, Medina gathered information from farmers, fellow *Andinistas,* and anyone who lived in the area and had an opinion on what the volcano might do. In early March, a town hall meeting was scheduled at the National University of Manizales. The scientists from *Piso Once* who were working on Nevado del Ruiz were to give a public seminar on the state of the volcano.

Marta Calvache volunteered to speak about the volcano's eruptive history and went to the National Historic Museum in Bogotá to find out what she could. At the museum's library, she pulled a dusty leather-bound volume from the shelf: *Comptes rendus,* the Academy of Sciences, Paris, 1846.

She opened up the book and turned its yellowed pages to the index, where she found reference to "the muddy eruption of the Ruiz volcano and the catastrophe that occurred in the Lagunillas River valley in the Republic of New Granada."

On page 709, she found a two-page account written by a Colombian naturalist named Joaquin Acosta. She read slowly, struggling with the French text.

> The 19th day of February in the year 1845, a great subterranean noise was heard in the vicinity of the Magdalena, from Ambalema to the village of Mendez. . . . This sudden noise was followed, within a considerably smaller area, by trembling of the ground. Then, descending along the Lagunillas from its sources in the Nevado del Ruiz, came an immense flood of thick mud which rapidly filled the bed of the river, covered or swept away the trees and houses, burying men and animals. The entire population

perished in the upper part and narrower parts of the
Lagunillas valley. In the sideways to the heights; less happily, others were stranded on the summits of small hills
from which it was impossible to save them before they
died. . . . On arriving at the plain with great impetus, the
current of mud divided into two branches. The much
larger one followed the course of the Lagunillas toward
the Magdalena; the other, after topping the high divide,
traversed the Santo Domingo valley . . . and hurled itself
into the Rio Sabandija, which was thus plugged by an
immense dam. The danger seemed imminent for a flood of
downstream lands. Happily, plentiful rain during the night
produced enough water to begin clearing a passage across
this mass of broken trees, sands, stones and stinking mud,
mixed with enormous blocks of ice which descended from
the mountains in such abundance that after several days
they had not melted entirely. . . . The terrain covered by
debris and mud is more than four leagues square; it pre-
sents the appearance of a desert or playa, on the surface
of which loom up like many islands heaps of broken
trees that resisted the impetus of the torrent. The depth of
the mud layer varied greatly and is much greater toward
the upper part of the deposit where it often reached 5
to 6 meters. A realistic calculation indicates more than
300 million tons of muddy material came from the flanks
of Volcán Ruiz.

Calvache pictured the summit of Nevado del Ruiz as she had
seen it on each visit, and the way the Lagunillas carved its way
through the mountains. She could easily visualize how the river
must have flowed with the horrifying avalanche of ice and mud.

She opened another book, *Noticias historiales de las conquistas
de tierra firma en las Indias Occidentales,* written by a Spanish
priest named Pedro Simon in 1625. What he reported was an event
that took place in 1595, 251 years earlier than the eruption and
consuming mudflows witnessed by Joaquin Acosta.

It happened, then, that on that day, month, and year (12 March 1595) . . . there came from this volcano such a loud, hoarse, and extraordinary thunderclap, and after it three others not so strong, which were heard within a radius of more than forty leagues. . . . The Spaniards saw that the volcano hurled out a large amount of pumice, as big as ostrich eggs and from these down to the size of dove's eggs, sparkling red like iron from the forge, which resembled erratic stars. Some fell on them and on their horses, which disquieted them no little. And on the side of this mountain which faced the east . . . the waters of the Rio Guali, which wets the foundations of Mariquita, it and its companion which flows in the south, called the Rio Lagunillas, both originating in the snow which melts from this mountain, ran so full of ash that it looked more like a thick soup of cinders than like water. Both overflowed their channels leaving the land over which they flowed so devastated that for many years afterward it produced nothing but weeds.

After several hours in the library, Calvache gathered her notes and went back to Manizales to work on the presentation she would give the next day at the university. It was March 18, 1985, nearly a century and a half since the last recorded eruption of Nevado del Ruiz.

While the volcano continued to rumble softly in April and May, a turf war began to brew between Bogotá and the Cafetera. Antagonistic feelings between the two regions were not new, and with the ensuing crisis, suspicion and distrust grew, and both sides demanded full control of Nevado del Ruiz. The most complicated aspect of the battle had to do with the seismic information collected from the volcano. It had been proven, just five years earlier when Mount Saint Helens erupted, that the data collected by seismographs is the most important tool for predicting an impending volcanic eruption. But at the time that Nevado del Ruiz started trembling, there were only six seismometers in the entire country of Colombia—the same network that had been installed by Father Jesús Emilio Ramírez in the 1960s. The seismographs recorded the many earthquakes that plagued the country but were too far away from Nevado del Ruiz to pick up the volcano's gentle tremors. The Jesuit scientists also wanted control of the volcano, but their experience in seismology was limited to earthquakes, not volcanoes; nor did they have the funding to come and work on Nevado del Ruiz.

The situation was no better at the National Institute of Geology and Mines, the national government agency headquartered in Bogotá that was in charge of geological studies in the country. They also had no money, no equipment, and no scientific expertise in the workings of an active volcano. However, the lack of knowledge didn't stop the agency from decreeing that it would be in charge of interpreting any and all seismic data collected from Nevado del Ruiz.

By late June, the government scientists, who came from Bogotá, Medellín, and Ibagué and refused to collaborate with Pablo Medina's group, had scrounged together three seismographs from a Colombian electric company. However, the instruments came without seismometers—the probe that goes into the ground to detect seismic movement—rendering them useless. The scientists had no idea what to do with the seismographs when they arrived, so they asked for help from the UN Disaster Relief Office and the United States. The U.S. Geological Survey (USGS) had become the world's expert agency in dealing with volcanic crises. The USGS was made up of government scientists who, unlike their Colombian counterparts, were officially in charge of all volcanic and earthquake emergencies in their country. The Survey brass agreed to send three seismometers, which would be paid for by the United Nations, but declined to send any technical assistance, claiming their volcanologists were tied up with volcanic activity at Mount Saint Helens and in Hawaii.

In August, the Colombian scientists' luck took a turn for the better. Bruno Martinelli, a tall, good-natured Swiss volcanologist, agreed to come to Manizales with three of his own seismographs. He was astonished to find that no one in the group of government scientists knew how to install or operate the equipment—or how to analyze the seismic data. Worse, the three instruments they had installed on the mountain were not working properly and were not positioned to be of much use.

"I became disgusted by the incompetence of Ingeominas [Colombia's National Institute of Geology and Mines]," says Martinelli. "I preferred to work with Bernardo Salazar and Juan Duarte." Martinelli found that Duarte, a technician in his forties from the Jesuit University in Bogotá, was the only person in the whole country who had any knowledge of what to do with a seismograph. But even Duarte didn't know how to interpret the signals from an active volcano.

Martinelli, Duarte, and Salazar had the job of installing Martinelli's three Swiss instruments and repositioning the others. The work was incredibly difficult. Martinelli had grown up climbing the Swiss Alps, but the *refugio* on Nevado del Ruiz was as high as the tallest peak in Switzerland, and it was another 2,000 feet to the volcano's summit. "The first two weeks were very hard," Martinelli remembers. "We tried two or three times to go to the crater. We had to carry the seismometers with us, and every fifty meters, we would stop." The freezing wind never seemed to quit blowing, and the thin air stank bitterly, burning their lungs.

Finally, in late August, the seismometers were in place, and Duarte, Martinelli, and Salazar began to make daily treks to the mountain to collect the data and change the drums.

But in reality, Martinelli didn't need new data to make the fundamental call. The crater of the volcano, once covered by ice, was now a deep, steaming pit, and earthquakes were rattling the mountain with increasing frequency: "I told the Colombian government the volcano is active and likely to erupt in the not-so-distant future," Martinelli says.

On the evening of September 10, Martinelli and Salazar drove back to Manizales. Salazar took the Swiss scientist to the Hotel las Colinas. The next morning, Martinelli rose late, made a visit to *Piso Once,* had a leisurely lunch, and went back to his room.

"I was preparing my luggage to go back to Switzerland when J.J. and Maria Elena called Bernardo Salazar by radio at 1:30. Bernardo called me at the hotel. He said that J.J. was saying it was raining ashes at the *refugio,* and that he felt a lot of quakes. He said J.J. sounded pretty serious. I thought maybe J.J. was exaggerating, but Bernardo thought we should go and see."

The day was overcast, and the two scientists couldn't see more than a half-mile ahead as they drove toward Nevado del Ruiz. It took two hours to make the trip. At 4 o'clock in the afternoon, they had reached the road that led up to the *refugio.* The whole sky was dark, as if it were a moonless night. Salazar turned on the car headlights. Ash began to fall. At first, it was just a little, like a light dusting of snow, but soon the ash was coming down so thick that Salazar had to inch the truck forward for fear of going off the road.

Bam!

A rock the size of a golf ball came out of nowhere and hit the truck; another followed. Soon it sounded like machine-gun fire hitting the roof. Flashes of lightning lit up the blackness with millisecond bursts of white. Salazar's instinct was to turn the car around and get the hell off the mountain, but he had to get J.J. Gutierrez and Maria Elena Vivas out of the *refugio.*

As he continued to drive slowly through the thickly falling ash, two figures emerged from the dark cloud. Gutierrez and Vivas, covered in dust and ash, staggered out of the blackness, holding their hands over their mouths. Vivas grabbed the rear door of the truck and jumped in. Gutierrez pushed the terrified woman from behind and climbed in after her. As quickly as he dared, Salazar carefully negotiated a three-point turn and drove down the dirt road through the murky rain of ash.

The national park was closed to visitors, and the *refugio* was off-limits to all but scientists working on the volcano. The eruption dropped less than a millimeter of ash onto Manizales, but the people could no longer ignore the live volcano looming above them. A bright, clear day followed the eruption, and a white plume of gas could be seen escaping from the steaming summit. The media went on a rampage to report the most exciting aspects of the eruption, and soon the whole country was aware that Nevado del Ruiz was no longer sleeping.

The following week, Marta Calvache came back to Colombia from a meeting of geothermal scientists in Hawaii. This time she was better equipped: "I had a pair of real mountaineering boots I bought in the States, and Pablo Medina had new climbing gear for all of us." For the first time, they would be hiking with crampons, ice axes, and other mountaineering essentials. "It was like we had uniforms," she says. "We even had matching jackets that said Comité de Estudios Vulcanicos across the back."

They hiked to the top in just under two hours. The crater was deeper than they had ever seen it, and the snow was covered with black ash. But the fumaroles were flowing less vigorously than before the eruption. There was no evidence that there had been pyroclastic flows, and from what they could see, no mudflows had been generated by glacial meltwater. The fact that the volcano had erupted gave Calvache a strange feeling. They had been on top of the volcano so many times, and their job had become almost routine. Even though she had studied the accounts of previous eruptions on Nevado del Ruiz, seeing the actual evidence of an eruption was unsettling. For the first time, she realized how incredibly dangerous and unpredictable the volcano was, and how they could easily be killed if they were standing on top of it.

Back in Manizales, the offices at *Piso Once* were bustling with scientists from the local universities, the press, and sporadic visits by local and national politicians. The eruption had brought Nevado del Ruiz to the attention of the world's volcanologists, and both the Colombian government and Pablo Median's "Committee" seized the opportunity to ask for more help.

Pete Hall, an American volcanologist and UN representative who ran the volcano observatory in Quito, Ecuador, came to Manizales representing the United Nations. It was his second trip—he had been to the city and the volcano in May. He was astounded. "None of the early recommendations I had made had been followed," Hall says. There was no real monitoring system in place, there were no evacuation procedures. There was still complete chaos among the different groups, infighting between Tolima and Caldas, and distrust between the Cafetera and the powers in Bogotá. To his dismay, Hall found that the seismographs collected from the mountain were *still* being sent to Bogotá, and that there hadn't been a single report back for over a month and a half.

Hall made it his first order of business to see that the seismographs would be analyzed in Manizales. He chose a young geophysicist named Fernando Muñoz, a professor at the University of Caldas, and paired him with Bernardo Salazar, teaching them what he could about how to read the signals recorded on the seismographs. But Hall was not an expert on volcanic seismicity. Muñoz remembers: "What Pete Hall was showing us was how to read earthquake signals, not volcanic signals. At the time, I had no idea there was any difference." When he had wrapped up their half-hour lesson, Hall told them they needed to get a *telemetered* seismic station installed on the volcano. "I figured that Bernardo knew what a telemetered seismic station was, and I didn't want either of them to know that I had no idea what Hall was talking about," says Muñoz.

"Okay," Hall said to the Muñoz, "Fernando, you will be in charge of analyzing the data for fifteen days, and Bernardo, you will be in charge of collecting the seismographs for fifteen days. Every two weeks you will switch roles. When you work in the field, you'll live at the house in Arbolito, and at the end of two weeks, you'll bring the seismographs back to Manizales."

They both nodded in agreement and then stood silently as Hall left the building.

"Okay," Salazar said when Hall was gone. "I'll go to the mountain first." Then he handed Muñoz a stack of seismographs that he had brought with him. "Here," he said. "Start with these."

Muñoz took out his notebook and read the notes he had taken during Pete Hall's brief lecture and quickly came to the conclusion that something was amiss with the seismometers. "There were all these signals. It didn't seem possible that they could mean anything. I figured the instruments must be broken," he says. Muñoz was sure there couldn't be so much activity on the volcano.

Several days later, John Tomblin, the director of the UN Disaster Relief Office, came to Manizales and visited *Piso Once*. Days earlier, Fernando Muñoz had finally learned that a telemetered seismic station is a seismometer that sits on a volcano and sends signals directly to a laboratory seismograph. The young Colombian saw his chance with the distinguished British scientist. "I pulled John Tomblin aside and asked him for a telemetered station, and he told me they were trying to get a few telemetered seismographs from the U.S. Geological Survey." That way, the scientists at *Piso Once* would know what the volcano was doing immediately, instead of two weeks later—or, as in the case with the data that went to Bogotá, not at all.

While the U.S. Geological Survey debated how many telemetered seismometers they could afford to send to Colombia, they decided to send one of their scientists, a geologist named Darrel Herd, who had worked in the area ten years earlier for his doctoral thesis.

Herd arrived in Colombia, and on September 23, he gave a talk in Manizales to a full house of locals, press, and government officials in the National University auditorium. The tall, refined American expert dispelled many of their fears, confidently assuring the gathering that the volcano would *not* hurt the populated regions, because any eruption would only affect an area 6 miles from the summit. The scientists from *Piso Once* who had worked so hard to warn the government and the people about the danger of Nevado del Ruiz were crushed. "Everyone was there to hear Darrel Herd's talk. A bunch of local people, the journalists, the politicians—everyone," Calvache says. The news of Herd's optimistic forecast was met with great relief when it appeared in the Bogotá newspaper *El Espectador* the following day.

Several days later, Father Rafael Goberna, the director of the

Instituto Geofísico at the Jesuit university in Bogotá, also assured
the population in the September 1985 issue of *Magazin 8 Dias* that
there was no imminent danger.

> Today, because of our monitoring, we can tell the entire
> population that nothing is happening on the volcano that
> threatens the inhabitants of the region. If and when it is
> necessary to declare an emergency, the Instituto Geofísico
> de los Andes will do it. Before would only alarm the popu-
> lation without reason.

Both Herd and Goberna's assessments of the situation were
music to the ears of the surrounding chambers of commerce, especially in the states of Caldas and Tolima. Civic leaders were irritated
by the inflammatory press reports and what they began to refer to as
"the volcanic terrorism" that was causing real estate prices to plum-
met. The mayor of Armero was especially glad to hear the report.
His was a lovely town with a thriving rural economy helped by a
hefty tourist trade. It wasn't good for business to have a volcano
poised to bury the small city.

The mayor was, however, very worried about a large dam that
had formed in the Rio Lagunillas from the debris of the September
eruption. A flood resulting from the dam breaking could cause a lot
of damage to his city. The national government denied his request to
come and drain the dam, and he lacked the financial resources to fix
it himself. His frustration was apparent in an October 8 interview
by *Consigna* magazine: "The new Emergency Committee does not
have the necessary information or financial resources to do anything
in the event of a catastrophe," he said. "For this reason, the people
have lost confidence in the veracity of the information and have
commended their fate to God."

In the United States, the U.S. Geological Survey finally acknowl-
edged the urgency of the Nevado del Ruiz crisis and decided to send

Dave Harlow and Randy White, two of their most experienced seis-mologists, to Colombia, along with what would have been the country's first telemetered seismometer. On November 6, 1985, Harlow and White packed their bags and got ready to leave their homes in Menlo Park, California, for Nevado del Ruiz. Then they turned on the news.

Charging the government with failing to implement political reforms, a brutal urban guerrilla group known as the M-19 had ambushed two guards and taken over the Palace of Justice in Bogotá. Several hours later, without warning, army tanks under the direction of President Betancur came crashing through the front doors, opening fire on the guerrillas and their hostages. More than 100 were killed on the spot by government troops—11 of them Supreme Court justices. Over 200 more people disappeared. That night, in an attempt to cover up the political disaster, all live news from the capital was cut off. The two national television stations broadcast a soccer game and the Miss Colombia pageant. "The next day, we got telegrams canceling the country clearance," Harlow recalls. The U.S. State Department had called off the trip.

Both the National Institute of Geology and Mines and the scientists from *Piso Once* mourned the loss of help and equipment from the U.S. Geological Survey. However, they had no choice but to carry on. So on Wednesday, November 12, Marta Calvache, Nestor García, and several others ascended the mountain. The weather was perfect, and the strong equatorial sun shone brightly on the summit while they collected samples from the fumaroles. There had been a fresh snow, and the mess made by the September eruption was covered by a smooth expanse of sparkling white.

Calvache's group returned to Manizales and met for their usual "state-of-the-volcano" meeting with other scientists who were working in *Piso Once*. "We told them that everything was normal on the summit," she says. Nothing seemed to have changed since the September eruption two months earlier.

"The next day in the afternoon it was really, really dark because there was so much rain, and I was glad we went to the volcano the

day before." From her office at *Piso Once,* Calvache occasionally
looked out of the window, searching for the volcano's summit, but
the rain continued and there was nothing to see. "I didn't know
what happened at Nevado del Ruiz until back at my apartment that
night, when I got a call from *Piso Once.*" Ash was falling in northern Tolima, her colleague told her.

Marta hung up the receiver and felt a rising sense of panic. It
could mean only one thing. The dreadful possibility, anticipated by
the local volcanologists and then dispelled by the calming voices of
a venerable priest and an American scientist, had finally become a
reality: Nevado del Ruiz was erupting.

Calvache turned on the television. A mudflow had hit Chinchiná,
the reporter said, and three people had already been killed running
from the deadly avalanche. She pictured the geographical layout of
the area surrounding Nevado del Ruiz. *If people were dying in
Chinchiná, what was happening in Armero?*

Calvache called Nestor García. They agreed to meet in the
morning to see for themselves what had happened at Nevado del
Ruiz and investigate the aftermath of the volcano's fury. Then she
crawled into bed, feeling helpless. Right now, in the dark, there was
absolutely nothing she could do but pray for the people in Armero
and Chinchiná.

CHAPTER 4: A HORRIBLE CONFLUENCE

NEVADO DEL RUIZ, COLOMBIA:
NOVEMBER 13, 1985, 9:30 P.M.

UPON RECEIVING Bernardo Salazar's frantic message from Arbolito at 9:15 P.M., the Risk Committee in Manizales immediately called the governor's office. Radio stations were soon issuing red alerts for the western communities of Caldas located along the rivers born from Nevado del Ruiz. Within minutes, the national television station was broadcasting news of the eruption.

In Armero, the city that lay along the Rio Lagunillas at the foot of the mountains 40 miles from the summit of Nevado del Ruiz, ash that had dirtied the town in the afternoon from the earlier eruption had quit falling, but a torrential rain continued, and businesses remained closed. There had been no word of the eruption called in by Salazar, and on the church loudspeaker, the village priest of Armero once again told the population to remain calm.

Juan José Restrepo, Sofía Navarro, and several other geology students who were part of the field trip group from Manizales listened to the radio in their hotel room in Armero. "The mayor was talking and he said not to worry, that it was a rain of ash, that they had not reported anything from the Nevado," Restrepo told U.S. Geological Survey scientist Jack Lockwood in an interview shortly after the eruption. If the ash starts to fall again, the radio announcer said, use a wet handkerchief over your face and avoid breathing the polluted air.

The time was 9:35 P.M.

On the summit of Nevado del Ruiz, heat from the eruption was melting millions of tons of the mountain's glacial cap. The new flows of hot rock, steam, and ash mixed violently with the glacial meltwater, forming a thick, hot fluid that rushed down along existing riverbeds. The warm debris-laden mud continued to thicken as it surged through the river valleys, ripping up soft, wet sediments several yards deep as it tore through steep channels. Walls of mud filled the steep valleys 60 feet above the riverbanks and raced down the canyons at more than 50 miles per hour.

The flows came in surges. Where there were bridges, the avalanche dammed up briefly, pausing as though catching its breath, until the terrible force of the deluge broke through, and the flood continued along its ruinous path.

The mudflow took its first victims at 10:30 P.M. in the Caldas village of Chinchiná, 15 miles due west from Nevado del Ruiz's summit. Dark and thick, unimaginably loud, it roared down the valley. Some residents living on the riverbanks had been evacuated, but the wave was much higher and moved more quickly than expected. Villagers scrambled desperately up the valley's steep slopes, but few could outrun the mudflow. In less than sixty seconds, more than 1,000 Chinchiná residents were dead.

Frantic radio transmissions went out to Manizales for help, and Civil Defense rescue workers, completely unprepared for what they would find, raced to Chinchiná in ambulances. To the east, two rivers—the Rio Azufrado and the Rio Lagunillas—became drains for the torrential mudflows barreling down from the high summit of Nevado del Ruiz. The crest had reached 100 feet high by the time it raced through the steep canyons of the Rio Azufrado, heading toward its juncture with the Rio Lagunillas.

Officials in Ibagué, the capital city of Tolima, had barely begun preparing to evacuate Armero's 30,000 residents when the entire city lost power. Telephones and radios quit working; but with the priest's reassurances freshly in mind, the townspeople calmly lit candles and prepared their children for bed.

By then, the avalanche traveling the Rio Azufrado toward Armero was just half an hour from its confluence with the Rio Lagunillas, where the mud already stood 30 feet above the riverbanks. Families of the small mountain villages that lined the Rio Azufrado came from their homes, shouting pointlessly, unable to hear or be heard above the thunderous roar. As the mud passed, the earth shook as if overrun by a gigantic cattle stampede.

Most of the mountainside residents lived high enough above the steep Rio Azufrado river valley and were spared by the mud. Several who had witnessed the flow were ham radio operators, and they immediately called out warnings to the low-lying city of Armero, just 20 miles down the river from their mountain homes. Some residents of Armero who received those calls escaped down the narrow, wet streets in small cars and trucks. Others scrambled awkwardly on foot, crippled by heavy bundles and dragging small children. Rain-drenched figures ran through the dark streets, illuminated sporadically by headlights of cars driven in a chaotic frenzy.

The people of Armero had been told that if Nevado del Ruiz erupted, they would have two hours to evacuate. But no one had ever told them where to go.

By 10:45 P.M.—more than ninety minutes after the eruption—firemen in Armero were finally alerted of the call to evacuate by Civil Defense. Without a formal plan and without power to sound an alarm, the firemen took to the streets blowing whistles and calling on their townspeople to come out of their homes and leave the city. Astonishingly few did. A young fireman became frantic. He went door to door, knocking, begging residents to leave their homes. Over and over he was told that the danger was a lie, that the padre had told them that there was nothing to fear.

Finally, fearing for his own life, the fireman gave up and left the city for safer ground.

After giving his original reassurances, the mayor of Armero left the local radio station and went back to his office on a dark street in the middle of town. Civil Defense in Ibagué was calling on the radio and warning that Armero must evacuate, but the mayor was tired and confused. He was much more worried about a flood from the dammed Rio Lagunillas. He spoke by radio to Civil Defense, arguing about the situation. "It's just an ash fall," he said. He wanted it all to be over, and he wanted to be with his family, who remained in their Armero home.

It was 10:50 P.M. when the noise came, quick and deep, a low rushing sound that grew louder by the second. From Ibagué, Civil Defense heard the mayor's last transmission: "Wait a minute," he called. "I think Armero is being flooded."

Across town, the geology students from Caldas University ran out to the street. "The cars were swaying and running people down," recalled Juan José Restrepo. "There was total darkness, the only light provided by cars . . . we were running and about to reach the corner when a river of water came down the streets. We turned toward the hotel and started screaming, because already the waters were dragging beds along, overturning cars, sweeping people away."

They ran to the hotel's third-story terrace and watched in horror as frantic drivers ploughed over the bodies of their neighbors, as children were ripped from the grasp of their mothers.

And then the mud came.

Restrepo's first thought was that the rumbling must be an earthquake. But the sound grew louder and louder. Suddenly, the mud appeared from the rear of the hotel like a wall of black foam coming out of the darkness. "Since the building was made of cement, I thought that it would resist, but it was coming in such an overwhelming way, like a wall of tractors, razing the city, razing everything." It crashed against the back of the building and began crushing the concrete walls. The ceiling slab fractured; within seconds the entire building was destroyed and broken to pieces, adding to the carnage of cars and refrigerators and bodies being carried away by the mud. Restrepo held on to a slab of concrete that had come to rest, and the mud continued to flow around him. Suddenly, a giant rectangular water tank was heading directly for him. He was sure the tank would run over him and push him into the mud to his death. Miraculously, the tank wedged against something in the darkness and came to a stop.

The mud, still warm, surrounded Restrepo as he continued to hold on to a piece of the stationary hotel wall. Restrepo looked up and saw the university school bus on top of the mudflow, higher than the hotel's mangled roof. The bus, on fire, exploded and sent glowing shrapnel into the sky. "I covered my face, thinking this is where I die a horrible death."

The moving avalanche of mud continued for several long, agonizing minutes. Restrepo heard screams for help throughout the darkness but was unable see where they came from. An old lady next to him was alive and struggling to breathe. She was covered with mud, her body wedged between two blocks of concrete and her face bloodied. Several feet away there was the body of a little girl who Restrepo thought had been decapitated.

"The old woman told me, 'Look, that girl moved a leg,'" Restrepo remembered.

Restrepo was sure the woman was hallucinating. He struggled to move toward the child but only sank deeper, his legs virtually immobilized. At last, he managed to reach her; her head was buried in the mud. He grabbed the girl by her shoulders and pulled desperately—harder and harder, but still he couldn't free her. Frantically, he plunged his hands into the mud around the girl's head, but it was hopeless. The child's hair was caught by something beneath the mire. Frozen waist-deep, Restrepo helplessly watched the small child go limp.

In shock, Restrepo struggled to stay alive as the heavy rain continued to pour. Then, suddenly, the ground began to rumble again. A second flow was coming, riding on the thick layer that had already buried the bodies of his friends and thousands of others.

Seconds seemed like hours to Restrepo. The wall that had

trapped him in the first deluge protected him in the second. Not so the old woman, who was swallowed by the moving avalanche. First her head and shoulders disappeared, then her torso, and finally her small, bare feet were pulled unmercifully into the hot mud.

For an hour, Restrepo listened to desperate screams from every direction. Then the land became silent, and Restrepo heard nothing more.

CHAPTER 5: THE MUD

ARMERO, COLOMBIA:
NOVEMBER 14, 1985, 7 A.M.

WHEN THE SUN ROSE on Armero, Juan José Restrepo squinted and looked at the barren landscape. He thought that it must be the end of the world. As his eyes adjusted to the sea of brown mud in the morning light, he soon recognized other slowly shifting forms, only the whites of their eyes visible. There were hundreds of bodies, completely covered in mud, some barely moving, some not at all. "We lost control when we saw that horrible sea of mud, which was so gigantic. . . . There were people buried, calling out, calling for help." It was a scene worse than any nightmare.

He wanted to rub the filth from his eyes, but his hands were completely covered with mud. Above him, an arm's reach away, rose the huge water tank that he had thought would kill him the night before when it came barreling toward the hotel. Now, the 5-foot-high, 30-foot-long concrete tank sat like a boxcar atop the mud. Restrepo thought that if he could reach the top of it, he would have a better chance of being rescued—at the very least, he would escape the mud, which was rapidly turning into a kind of thick cement around his lower body.

As Restrepo tried to work his way toward the tank, he heard a soft cry coming from behind him. It was Sofía Navarro. The last time he had seen her was on the hotel terrace; he had been holding on to her hand when the mud came and had thought she'd surely died. It seemed like a miracle that she had survived, that either of them had made it through the night.

He grabbed Navarro by the wrist to pull her closer to the water tank, but the mud refused to loosen its grip on her. Restrepo feared he would pull her arm from its socket, but Navarro didn't cry out. Eventually she got loose; Restrepo boosted her up to the top of the tank, then dislodged himself from the mud and climbed up as well. The view from above was unreal.

The beautiful, tranquil city of Armero no longer existed.

In its place was an endless expanse of wet, brown earth littered with the regurgitated remains of the once-quaint town. Cars and trucks were turned on their sides or upside down. The wood that had built the town's simple homes was scattered like straw. Horses and cows and pet dogs lay on their sides, stiff-legged. Most terrifying, they were surrounded by hundreds of weakened survivors buried up to their necks. Hundreds of disfigured corpses, too many to count, were scattered like discarded dolls. Some of the living had managed to climb into a stand of trees that had somehow managed to withstand the avalanche.

Several small hills rose from the mud like islands dotting a flat, brown ocean. A few houses had survived. Restrepo watched as hundreds of human forms crawled toward the hilltops, trying desperately to reach high ground. He lowered his head and sobbed, joining the sorrowful chorus of the others who had made their way onto the water tank. But he was jarred momentarily from his grief by a bloodcurdling yell.

A man lay helpless on the mud, his legs completely severed above his knees. Restrepo and another man lay on their stomachs and reached down to grab the man's forearms. "We helped the amputated man out of the mud and lifted him up onto a tank." Then they covered his amputations with mud so he would not bleed to death.

At 8 A.M., clouds obscured the sun, but the air turned hot and humid. The water tank now held a dozen injured people who had managed to get to the top for safety. A 5-year-old girl, who had been lifted to the tank by other survivors, rubbed her eyes with dirty fists as she cried. Her little nightgown was stiff with dried mud, and her legs were covered with cuts. That she had survived was a miracle. With tiny bare feet, she moved to the edge of the tank and looked down into the mire and called for her mother.

A woman with a large wound across her cheek went to the child, held the tiny girl, and tried to comfort her, to tell her that her mother would be coming soon. But the woman knew she was lying to the child. The child's mother most likely would *not* be coming any time soon. The night before, the woman's own son had been torn from her grasp as a current of wet earth carried him away into the dark.

Restrepo heard a plane flying overhead. "I took off my shirt and waved it around," he recalled. Several of the men joined him, but to no avail. "It never saw us." Restrepo fell to his knees, and for a full hour he prayed—in vain—for the plane to return.

The surface of the mud was littered with food from open refrigerators that had been carried by the avalanche and had spilled their contents. Restrepo and several others began to gather fruit and sodas that were within reach.

On the tank, the injured were growing increasingly frightened, and Restrepo did what he could to try and calm them. The young geology student felt that he had to remain in control, because he seemed to be the only one who had made it through the night without injuries. There was no water to drink or supplies to help tend to the wounded. One man's eye was gouged out; a deep red-brown hole was all that remained. Another's head looked like it had split, and his thick hair was soaked with sticky brown blood. Sofía Navarro had a 6-inch gash on her thigh, and she could only walk very slowly. Mostly, she sat quietly. Restrepo said a prayer and, standing in the center of the tank, tried to encourage the group around him. He

remembered, "I told them, 'They're going to save us. If we aren't rescued today, then God will make sure we are rescued tomorrow.' "

Just before 3 o'clock that morning, Marta Calvache was awakened
by the sharp ring of the telephone. It was Nestor García. He
sounded panicked. "The eruption killed thousands of people in
Chinchiná," he said. Calvache was instantly awake and sitting up in
her bed. "We'll be by to get you in fifteen minutes."

She put the phone down, hurried out of bed and into her clothes,
and hastily grabbed the gear she had packed the night before.

García and Bis arrived at her apartment, and with several others,
they drove to the mountain to see what damage had been done. By 6
in the morning, they came to the small house in Arbolito, where
Bernardo Salazar and Fernando Gil had witnessed the eruption the
night before. They stopped and stood alongside the dirt road and
looked to the southeast. Dawn offered the first view of the volcano's
summit.

"We saw pumice everywhere around the volcano, everything was
gray," Calvache says. The scientists stood silently and stared at the
volcano. The massive glaciers that had looked like giant fangs coming down the valleys were half their original size, and the snow on
the summit was black with ash. Along the steep riverbed of the Rio
Guali, the mud had careened through the river channel, leaving a
scar 100 feet high where trees and soil had been ripped from the
sides of the valley.

A minute later, they were back in the truck and driving toward
the mountain. The road was covered with a thin coat of marble-
sized pumice pebbles that crunched under the car's tires. When they
reached Rio Guali, the bridge that crossed the river had completely
vanished. The driver stopped the truck at the steep cliff, and the scientists got out and surveyed the destruction. The river, usually only
a narrow path of icy water, was a thick bed of coffee-colored mud,
and a small trickle of fresh water had just begun to etch a sinuous
path into the wet earth. "The first thing we wanted to do was try to

get close to the *refugio*," Calvache says, "and recover the [seismic] records. But all the roads were cut." With the bridge gone, there was no way to drive to the *refugio*. They got out of the car, and Bis opened up a map. With his index finger, he pointed out an alternate route up the mountain. The group began to hike, but the mountains were steep and covered with thick vegetation, and their progress was slow. They hiked for several hours, and then it began to rain.

Marta Calvache's mind was reeling. After seeing the Rio Guali, she could only imagine the damage done on the other side of the mountain, where the flow from the Lagunillas would have disgorged onto Armero.

At 8 A.M. the radio crackled. It was *Piso Once*. "There is a report from a pilot who flew over Armero this morning. The report is that the city no longer exists. It is completely covered in mud."

García and Calvache stood frozen and stared at the ground. Their most horrible nightmare had come true.

In the afternoon, Juan José Restrepo and the group on the water tank had grown very quiet. There was nothing left to drink or eat from what they had scavenged, and the day was muggy and stiflingly hot.

From far away, Restrepo heard a sound. At first, he thought he was imagining it, but the faint, rhythmic sound grew louder and more distinct: a helicopter. As the wounded survivors screamed for help, the helicopter finally appeared from behind the clouds: a small, military aircraft seemed to be coming directly toward them. But the aircraft hovered over a nearby hill and dropped a rope ladder that danced violently under the gusty wind from the rotor. In an instant, a rescue worker in a bright orange jumpsuit climbed down the ladder and began to fasten a harness around an injured child. Immediately, he was surrounded. The people on the hill screamed, trying to be heard over the clamor of the helicopter.

Restrepo and the others on the tank yelled desperately as the rescue worker fastened harnesses around the injured, and then they became silent as the victims were lifted to safety. The aircraft disappeared into the hazy sky. Restrepo's heart sank. He thought of his family back home in Manizales. They would have no idea he was here in Armero in the middle of this horrible nightmare. He was supposed to have been in Ibagué with the rest of the geology students. He thought about his friends and their families and wondered if any more of his *compañeros* were alive.

The day grew hotter and hotter. Hours went by, and the helicopter didn't return.

Earlier in the morning, Fernando Rivera, the crop-dusting pilot who had flown over Armero, had reported to the Tolima Civil Defense that Armero no longer existed. Rivera reported that he did not see any survivors—only a sea of endless brown filth. Red Cross and Civil Defense workers hurried to the small town with rescue supplies and gear but were completely unprepared for what they would encounter. It was impossible to get near the city. Mud covered all of the roads, and the bridges were destroyed. The rescue workers could, however, see hundreds of people who were still alive. It was immediately obvious why the pilot hadn't seen them from the air; the victims were camouflaged by the thick mud that covered their bodies.

The media from Ibagué followed on the heels of the rescue workers, and news of the devastation was soon on Colombian television. Hours later, news had spread around the world. The images were surreal; graphic scenes of Civil Defense workers stacking bodies like cordwood. A helicopter view showed hundreds of casualties stuck in the mud and screaming for help. The church steeple seemed to float above the mire. And at the local cemetery, perched on a small hill in the northern part of town, neatly tended plots and carved headstones stood completely intact.

In the United States, when the news reached radio and television in the early afternoon, very few people had ever heard of Nevado del

Ruiz. When Norm Banks, a volcanologist in his mid-forties who worked at the Cascades Volcano Observatory in Vancouver, Washington, got the news in midafternoon, he was horrified by the casualty report. He remembered hearing about activity at a volcano in Colombia several months before. He had been told that the U.S. Geological Survey was planning on sending some equipment but wouldn't be sending anyone for technical support.

For Banks it was a particular blow. He had spent more than four years—since the eruption of Mount Saint Helens—trying to establish the world's first mobile volcano observatory.

It had been an endless uphill battle. Each year he would write a proposal to the survey, and each year it would be turned down. Meanwhile, he stockpiled what spare instruments he could and waited for the call he was sure he'd get. It came in the summer of 1981. A small volcano named Pagan in the northern Mariana Islands had erupted. It was no secret that Norm Banks's office was full of monitoring equipment, so the decision was made to send him to Pagan.

Banks was able to accurately forecast the next eruption on Pagan, and he became a local hero on the island. He took the opportunity to convince the local government of Pagan that they would be well served by the monitoring equipment that he had trained the islanders to use. And instead of hauling back his scavenged parts, Banks came back with enough money to replace everything he left at Pagan; he now had six trunks of shiny new machinery, which he then took with him to the next eruption—this time in Indonesia. After that, Banks took what hadn't been stolen by Indonesian customs agents to Papua New Guinea, where another volcano, named Rubal, had started to rumble. As before, he collected enough money to buy even more new gear, leaving the original equipment in place. Banks repeated the method for another four successfully managed volcano crises around the world, and ended up very well stocked. "Norm was pouring his heart into this because he knew it was a critical need," says Jack Lockwood. Despite all of Banks's success, it wasn't until 1985 that the USGS brass would finally make the

Volcano Crisis Assistance Team, or VCAT, an official program—
and even then, the USGS refused to make the VCAT available to the
scientists working at Nevado del Ruiz prior to the eruption.

The helicopter never returned. As night fell, Juan José Restrepo and
the others on the tank were besieged by swarms of mosquitoes and
biting flies. Some people died in the night. Some lay awake with the
pain of their injuries. They were very thirsty and painfully hungry.
Sofía Navarro's infection was worsening.

Throughout the night, Restrepo would fall asleep for several
minutes and then wake to a horrifying nightmare that the volcano
was erupting and the mud was coming. The night lasted an eternity,
and when the sun rose again, fewer of the bodies surrounding the
tank were moving.

A rescue seemed unlikely, considering the thousands of victims,
and Restrepo decided to take action. He proposed that the stronger
people build a bridge across the mud that would reach a small hill
two blocks away. Restrepo remembered, "We took some tables and
everything we could find in the mud and made a bridge." But the
overpass could support only one person at a time and was a perilous
crossing. "If you were walking on the bridge and you fell in the mud,
you would be stuck there and die." The hill they were trying to reach
rose a mere 50 feet in elevation above the city, but there were houses
still standing on its narrow summit. While they worked, they occasionally heard helicopters, but none landed nearby.

There were sixteen of them left alive on the tank, and as they
began their journey across the bridge, they passed dozens of bodies
that had begun to rot in the heat. Restrepo remembered: "Everything smelled ugly. Around us, there were a lot of dead people and
mosquitoes and flies, and I had to plug my nose to breathe." At last
they reached dry ground on the hill and went inside one of the three
houses that still stood. The house looked like the residents had left
in a hurry. There were open drawers with clothes falling out, there
were family pictures knocked over on the bedroom dresser. "The

people were becoming more panicked because they were hungry," Restrepo remembered. "We ate some raw eggs and drank Coke." Then they raided the closets and changed out of their filthy clothing.

Some of them left the first house and went to the two other houses to look for food and water. Later, they sat in the street and spent the afternoon watching the helicopters rescue victims as more survivors crawled silently from the mud onto the hill and joined them.

That afternoon, on the other side of the Cordillera Central, the scientists who had worked to prevent a disaster at Nevado del Ruiz gathered at *Piso Once*. The mood was somber. *Piso Once* was now a tense twenty-four-hour operation. For the past two days, there had been nonstop meetings with the local and national government, international scientists, and the press. "We had so much work to do, it was crazy," says Fernando Muñoz. "We'd work forty hours straight and then sleep on the tables." It was November 15. There were over 1,000 reported dead in Chinchiná, more than 20,000 in Armero, and the body count continued to rise. A white fountain of steam that stretched from the summit of Nevado del Ruiz 2 miles into the atmosphere was visible from Manizales, and the scientists had no idea if the volcano would blow again.

Pablo Medina sat at a large conference table, his unshaven chin cradled in his palm, his white shirt rumpled and rolled up at the sleeves. He was surrounded by dozens of scientists, among them Bernardo Salazar. Salazar had seen the eruption from the house in Arbolito, he had heard the mudflows, and he had called by radio to warn that they were coming. He had told them over and over and over. *Had he not been clear enough? Why hadn't Armero evacuated?*

Fernando Muñoz, the tall, clean-cut geophysics professor who had been working to interpret the seismographs with Bernardo Salazar, was also heartsick. Everyone *knew* that Armero was at risk—they had drawn hazard maps that showed the exact path of the mudflows. "The eruption changed our lives, everyone involved. It was too sad because we had the information. Bernardo was really bad. He had the information that night in the field. We could see in his eyes that he was destroyed, and he didn't want to talk to us."

Marta Calvache was similarly affected. "We had been to the summit the day before, and there was absolutely no reason to think the volcano was going to erupt. Everything looked normal."

Suddenly, foreign scientists who had previously declined the invitation to come to Nevado del Ruiz were now bringing state-of-the-art equipment to Manizales. The first to arrive at *Piso Once* were Bruno Martinelli, the big Swiss seismologist, and John Tomblin, the director of the UN Disaster Relief Office.

Tomblin arrived the day after the eruption from Hawaii, where he had been working at the Hawaii Volcano Observatory. "He brought us one unofficially 'borrowed' telemetered seismograph," recalls Fernando Muñoz.

Muñoz looked at the instrument with its poles and antenna that would transmit seismic signals back via microwave radiation to a seismograph in *Piso Once*. It was the first telemetered seismograph that had ever been in Colombia. *A little late,* he thought.

He did not blame John Tomblin. In fact, he couldn't point the finger at any one person or entity. Yet the fact remained that a handful of young Colombian scientists with no idea what to do had been left to manage the crisis. With almost no support from their own government or the rest of the world, they had given their hearts and souls to preventing a disaster. But they were virtually helpless. The volcano erupted, and more than 23,000 people were dead.

Juan José Restrepo and the others remained on the small hill for the entire night of November 15. Restrepo hadn't slept for two days and still couldn't. He spent his time caring for Sofía Navarro and the other wounded. By late afternoon, the helicopters had quit coming, and in one of the houses, Restrepo found a portable radio and turned it on. "We heard on the radio that everyone had been rescued," he remembered. There were groans of despair from the dozens of people who had found their way to the hilltop.

The hours went by slowly, and Restrepo lost track of time. He found himself wondering how long it had been since the mud had come. He thought again of his family. They would wonder why he hadn't returned from his field trip.

Restrepo remembers: "We watched the helicopters take more people out. We had to stay there all night taking care of other people. We couldn't sleep. Then we heard some dogs barking, and the people thought that it was going to happen again. The radio said that the Rio Lagunillas was rising again. I felt like death was breathing down my neck."

Hours went by, and the mud didn't come. The air reeked with the stench of rotting corpses, and Restrepo and the others on the hill plugged their noses and prayed that they would live through the night.

When the sun rose, the mud had at last dried enough so that it could support their weight and they began to walk north toward the cemetery, which was now covered with victims waiting to be rescued. The swampy land was still soft and difficult to walk through, and most of the people were injured and moved very slowly. Finally, they came to the hill where stone crosses and etched headstones remained untouched by the mud.

Several helicopters came and dropped water and food that was instantly devoured by the starving victims. Restrepo had little hope that they would be rescued from here either, with the swarms of victims that were now crawling to the hilltop graveyard. They rested for a while and then continued walking. Navarro was limping badly, and their pace was painfully slow. They eventually reached Guayabal, a small village several miles to the north. The town was crammed with wretched-looking people. It seemed to Restrepo that all of Colombia had come to Guayabal. People in cars were handing out milk and water and bars of sweet brown sugar. Hospital tents were set up all along the road, and hundreds of survivors walked like zombies through the streets, stopping people who passed them to ask about loved ones. "A lot of people came to try to find family and friends," Restrepo remembered. "We just tried to give hope to the people; we told them not to worry."

Restrepo knew he would live, but the horror was so great that he felt no relief, just complete and utter exhaustion. He felt like the volcano had ripped out his soul.

The roads and bridges that crossed the Cordillera Central were washed out for many miles to the north, and it took Juan José Restrepo and Sofía Navarro two more days to travel by bus to the other side of the mountain range. In Manizales, the families of the Caldas University students had gathered to pray for the safe return of their children. There was great elation as the exhausted students slowly found their way back to Manizales, alone or in small groups, until the count reached nineteen survivors. But the mood was short-lived: The professor, Jorge Dorado, the bus driver, and ten of the geology students would never return from Armero.

CHAPTER 6: THE INVASION

IT HAD BEEN FIVE DAYS since the eruption of Nevado del Ruiz. The rescue effort in Armero was now just a cleanup operation, and *Piso Once* was completely overrun by scientists from around the world. There were Italians, Japanese, French, Swiss, Spanish, and more than a dozen U.S. scientists.

Marta Calvache walked into the eleventh floor of the Banco Cafetero building on Monday, November 17, and didn't recognize a single person there. "It was difficult to even *find* a Colombian," she says. There were *gringos* everywhere. They were setting up big computers and talking loudly over maps spread out on the large conference table. Calvache eventually found Pablo Medina, who was trying his best to keep the operation calm and organized. He offered to pair a Colombian scientist with each of the foreign groups to help them navigate their way around Nevado del Ruiz, and Bis was available to guide the international scientists up the volcano.

Bernardo Salazar and Fernando Gil accompanied U.S. scientists to Nevado del Ruiz, where they installed the six telemetered seismographs that the Americans had brought with them. Fernando Muñoz stayed at *Piso Once* and worked with Dave Harlow, the same seismologist who had been packing to come to Manizales just a week before the eruption but whose trip had been canceled after the Palace of Justice tragedy. "It was chaotic—scientists and government officials all trying to come together to see if the volcano was going to erupt again," says Harlow. Harlow explained to Muñoz how to use the 40-pound computer that the Americans had lugged with them. With data received directly from the telemetered seismometers, the computer would process the information and help the scientists determine where, within the volcano, the fractures or tremors were coming from. Muñoz was a quick study, and the computer was much faster than anything he had ever worked with before. Next, Harlow sat down with the young Colombian and began to explain what the seismic signals that came from an active volcano looked like. "I looked at all of the records and tried to see if we could find similar patterns, indicating that another eruption was coming," Harlow says.

Muñoz was astounded. All of the signals on the seismographs that he had attributed to problems with the instrument or wind had actually been signs that Nevado del Ruiz was getting ready to blow. Muñoz wondered what might have been done differently if Dave Harlow had been allowed to come *before* the eruption, as originally planned.

Norm Banks had also received the okay to bring the Volcano Crisis Assistance Team to Colombia. He arrived in Manizales with a jet full of equipment and two assistants and was put in charge of the U.S. Geological Survey group of scientists, while Darrel Herd was sent to be the liaison between the USGS and the government in Bogotá. Banks soon found out that Herd had been in Manizales before the eruption. At first Banks was stunned—then furious. His bosses had told him that they wouldn't be sending anyone. He didn't want to believe that the Geological Survey hadn't sent his own VCAT, or at least someone better trained in handling volcanic crises. "Darrel was there, back when he did his thesis on the glaciology of

Ruiz, but he had no experience dealing with the press and public. His intentions were good, but he had no idea what to do," says Banks.

Only later did Banks learn of Herd's statements to the Colombians that the volcano would not harm the surrounding populated areas.

While the Colombian scientists tried to get a handle on who was who, personality conflicts and power struggles ran rampant among the foreigners. Haroun Tazieff, a French volcanologist who also happened to be a high-ranking politician, arrived with a pack of French press and became one of *Piso Once*'s biggest headaches. He immediately commandeered a U.S. Black Hawk helicopter (one of several that had been brought to Manizales after the rescue effort in Armero was called off) and demanded that all of the French television journalists accompany him to the volcano. "It was a crazy time," says Bruno Martinelli. "All the journalists wanted to go in the helicopters, but the scientists told them that only the people working on the volcano could go. So now all the Colombian journalists are complaining that the French journalists have helicopter access."

The situation became even more ridiculous when Tazieff's secretary demanded that both Martinelli and Marta Calvache accompany Tazieff in his helicopter to the summit of the volcano. "When we arrived at the airport," Calvache says, "the secretary told us that we were not allowed to talk directly to Tazieff. If we had anything to say, we would have to talk to one of his aides."

When the Black Hawk finally reached the summit, "Tazieff was yelling at the pilot to land in the volcano's crater," Bruno Martinelli says. "It was completely insane. I thought, 'This man is seventy years old and he's going to die in this volcano, and I'm going to die, too, but I'm not seventy years old yet.'" The pilot ignored Tazieff and put the Black Hawk down on the summit, a good distance from the crater.

In addition to Tazieff, Manizales was reeling with swarms of dignitaries who arrived to huge fanfare and conducted their benevolent visits with endless pomp and circumstance. The queen of Spain and planeloads of Spanish correspondents arrived to red carpets in Manizales and were soon followed by the wife of French president François Mitterrand, with similar revelry.

On an international level, the fire that President Belisario Betancur had been under for the Palace of Justice debacle was soon forgotten. Betancur arrived in Manizales three days after the eruption and gave Pablo Medina's group in *Piso Once* temporary control over decisions regarding the volcano, complete with a red telephone directly linked to the presidential palace.

Within a month, it was apparent that the volcano was quieting down. The monitoring equipment was in place, and the Colombian scientists had been trained in how to use the gear and interpret the data. After a not-so-friendly struggle between Bogotá and the Cafetera, President Betancur put the National Institute of Geology and Mines in charge of the volcano, and they went about hiring all of the young scientists who had worked on Nevado del Ruiz. Marta Calvache, Fernando Gil, and Fernando Muñoz went to work for the institute. The hard feelings between scientists that had festered before the eruption quickly faded at the newly formed Volcano Observatory. Nestor García chose to accept a job with the University of Caldas, teaching chemistry and martial arts. Bernardo Salazar walked out of *Piso Once* and never worked on Nevado del Ruiz again. Pablo Medina resumed his post at the Financial Corporation of Caldas. Bruno Martinelli left for his home in Switzerland but swore he would soon return to Colombia—and to Marta Calvache.

Things began to have a semi-stable feel in Manizales, but the insanity hadn't totally subsided. The international group of scientists had left the Colombians with conflicting prognoses of the volcano's future. Most were in agreement that the volcano was quieting down; there was one exception—a young American chemist named Stanley Williams. Williams was an assistant professor at Louisiana State University and virtually unknown to the U.S. Geological Survey volcanologists. He had come onto the scene days after the eruption and gathered gas samples from the volcano. The gas, he found, contained large amounts of sulfur, leading him to draw the terrifying conclusion that a catastrophic eruption was imminent and would make the earlier tragedy pale by comparison. "Stan was on a huge crusade that a monstrous caldera was going to form," says Dave Harlow.

The Colombian scientists asked Norm Banks to talk to Williams about his dire forecast, which was terrifying the locals. They also hoped Banks would be able to reassure the people of the region that the volcano was indeed quieting down.

Banks found Williams in *Piso Once* and asked if they could talk in one of the small offices off the main conference room. He let Williams explain his theory, and when he was finished, Banks told Williams that he didn't think the high levels of sulfur meant that a huge eruption was on the way.

"I told him that there would have to be a gargantuan ocean of magma rising up to the surface at a rate of several meters per day to produce that much sulfur—a complete impossibility," says Banks.

The reaction of the bespectacled, seemingly mild-mannered scientist from Louisiana startled Banks. "He was standing and yelling that I didn't know what the hell I was talking about. I told him that there were *alternative* explanations that he should consider before making claims that can cause a lot of fear—not to mention financial damage to the area." Williams and Banks continued what became a heated debate for another ten minutes; then Williams stormed out, slamming the door behind him.

Because the U.S. Geological Survey was the "sanctioned" scientific group called on by foreign countries, university scientists often felt like they were pushed aside in international crisis situations. Banks knew this, but he had never encountered such a degree of resentment before. The USGS credo was to support the locals and stay behind the scenes when they worked in foreign countries during crisis situations. It was a rule that Stanley Williams didn't seem to want to abide by.

Nevado del Ruiz never had the murderous explosion that

Williams predicted. Over the next several years, Williams would go on to make a name for himself as a notorious maverick who seemed to be constantly at odds with whatever the scientific consensus was. Williams believed he could predict volcanic eruptions by taking samples of gases from a volcano's crater; he seemed disinclined to work with seismologists or geophysicists—or, indeed, anyone who wasn't convinced of the predictive powers of volcanic gas. "Stan was still very dismissive of seismology and a lot of the stuff in some ways that was cutting edge," says Dave Harlow. This created a vicious circle: Williams would ask to be included in U.S. Geological Survey projects (and receive survey funds); the survey scientists, wary of Williams's uncooperative reputation, would tell him that they didn't need his expertise. Generally, the U.S. Geological Survey would be called to a volcano that was already in a heightened state of activity. They didn't need anyone going into the crater to tell them what they already knew, nor did they want someone so prone to cataclysmic pronouncements. But Williams would find the money and show up anyway, more irritated and inflexible with each new eruption.

The more generous members of the U.S. Geological Survey put down Stanley Williams's outbursts to youthful indiscretions of an inexperienced scientist in his thirties. And so, when Williams organized a workshop in Manizales in April 1988 (more than two years after the tragedy of Armero) and invited everyone who had worked on Nevado del Ruiz to attend, his peers hoped it was a sign of a new maturity. The workshop's point, as Williams wrote in the volume of scientific abstracts that would be presented at the meeting, was the need for more collaboration.

> I became convinced that there was a genuine need for a workshop while I was in Colombia during December 1986 and January 1987. During that long stay, I began to finally realize how much we have actually learned about Ruiz since those early days in November 1985. It may well rank as one of the best-studied volcanoes in the world.

Bis the *Andinista* on the summit of Nevado del Ruiz before the
November 13, 1985, eruption *(Photo credit: Bruno Martinelli)*

Marta Calvache and CHEC colleague taking a helicopter to inspect the
damage caused by the November 13, 1985, eruption of Nevado del Ruiz
(Photo credit: Arlene Collins)

Bernardo Salazar with a seismograph drum, Arbolito, Colombia, November 1985 *(Photo credit: Arlene Collins)*

Pablo Medina *(left)* and colleagues from Ingeominas and CHEC at *Piso Once* after the November 13, 1985, eruption *(Photo credit: Arlene Collinsz)*

Fernando Gil, November 1985 *(Photo credit: Arlene Collins)*

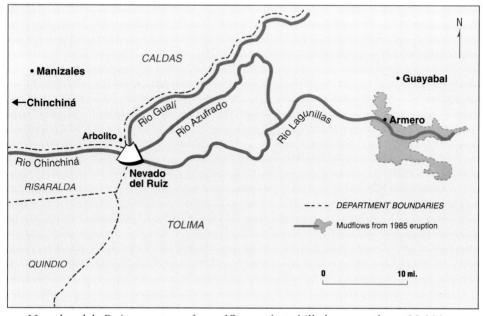

Nevado del Ruiz: routes of mudflows that killed more than 23,000 people following the eruption of Ruiz on November 13, 1985 *(Map credit: Sterling Spangler)*

The caldera and the cone of Volcán Galeras in 1994. View is from the southeast and shows the police station that was destroyed in the eruption in June 1993. *(Photo credit: Observatorio Vulcanológico de Pasto)*

The crater of Galeras from the southwest and the three fountains of steam that make up *Deformes Fumarolas (Photo credit: Observatorio Vulcanológico de Pasto)*

December 1992. José Arles Zapata (wearing a hard hat) tests the temperature of the fumaroles at Galeras. *(Photo credit: Bruno Martinelli, Observatorio Vulcanológico de Pasto)*

The Los Alamos team (in hard hats) with members of the Hotel Cuellar staff. *From left:* Alfredo Roldán, Fraser Goff, Andy Adams, Gary McMurtry. *Top left:* waiter/bellhop Luis Ruales. *(Photo credit: Alfredo Roldán)*

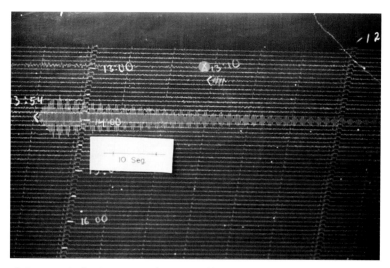

A long-period seismic signal, or *tornillo*, recorded at Galeras (*Photo credit: Bruno Martinelli, Observatorio Vulcanológico de Pasto*)

Approximately two hours prior to the eruption of Galeras. *Top, from left:* Alfredo Roldán, Stanley Williams, Nestor García, José Arles Zapata; *bottom, from left:* Fabio García, Igor Menyailov (*Photo credit: Fabio García*)

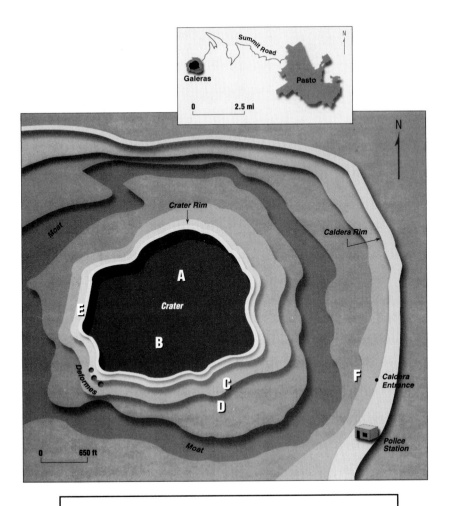

Galeras: Approximate position of scientists and tourists at the time of the January 14, 1993, eruption *(Map credit: Sterling Spangler)*

Perhaps more important in my decision to take on what has been a difficult task was the realization that some of the ideas, hypotheses, models, etc. that were slowly developing in the minds of various workers were usually based on only one of the many different sets [kinds] of data. I surely realized that my understanding of the deformation or seismicity was extremely limited and suffered from a lack of coherence. . . .

The point is, we are faced with a situation which is potentially extremely exciting exactly because of the apparent dichotomies in our different data sets. If we can actually sit down, educate each other, listen to each other, and ask the right questions, we may be able to fundamentally improve our conceptual understanding of the nature of activity at typical circumpacific stratovolcanoes. Ruiz has developed into a fabulous natural laboratory for our investigations. It is the time to reflect on what we have learned so far and plan where we will place our priorities in the future. The second worst volcanic disaster of the 20th century surely deserves our best efforts, both as individuals and as a collaborative team.

While many in the scientific community applauded this call for a cooperative, interdisciplinary approach to volcano research, others had difficulty reconciling this with Stanley Williams's usual independent style. The U.S. Geological Survey scientists were scratching their heads. Was Stanley Williams really ready to be a team player?

PART II: GALERAS

CHAPTER 7:
URCUNINA, THE FIRE MOUNTAIN

PASTO, COLOMBIA:
APRIL 1988

NEVADO DEL RUIZ WENT BACK TO SLEEP, but its reputation was now that of a pernicious killer rather than a peaceful giant. By the late 1980s, Colombia was troubled by deep economic depression; endless violence from guerrillas raged in the countryside; judges, police, and military officers were being systematically executed by the drug cartels; and another volcano was about to make a name for itself.

It was barely perceptible at first—just a thin stream of white steam emanating from a tiny pinprick. It came from the western flank of the cone, rose a few feet, and then disappeared. It was only a whisper, but there was no denying what it meant. The long-slumbering Volcán Galeras was beginning to wake.

It was April 1988, and for the past decade, military police had been stationed on top of Galeras, a 14,000-foot volcano that loomed above the city of Pasto in southern Colombia. The guards were there, in a cement hut on a narrow ridge of brown-gray rock, to keep watch on six communication towers built onto the highest point of the volcano's rock-strewn summit. While *La Violencia* was growing increasingly more frequent, the department of Nariño, where Galeras resided, remained mostly calm. But there was always some threat from the guerrillas.

Several years earlier, just 100 miles to the north along the Cordillera Central, rebel forces had climbed Volcán Puracé, killed the guards at the military outpost there, and destroyed the radio towers on the summit. Since then, the guards at Galeras had taken extra precautions. There were two ways to get to the transmitters. From the southeast, a rocky switchback road led all the way from the bottom of the mountain to the summit, but it dead-ended at the police station, where guards were on duty twenty-four hours a day. The only other way to reach the towers was to ascend the steep slope on the volcano's western flank, a crumbling, 30-degree incline prone to frequent rock slides. To deter anyone from ambushing the equipment from the rear, the military had laced the way with land mines.

The threat of violence from their fellow countrymen was very real, but the young officers had never been concerned about the volcano. For as long as they had been holding vigil—in fact, for as long as any of them had been alive—Galeras had been quiet.

As the wispy column of steam grew taller and poured more vigorously from the volcano, the guards began to take an interest in Galeras. The police station was located on the southeast ridge of the volcano's crescent-shaped summit. From the front of the building on a clear day, the guards had a perfect view of Galeras's newly active cone. It sat in a carved-out bowl and rose 450 feet from the floor of a mile-wide hollowed-out crater that scientists called a *caldera* (Spanish for "cauldron"). The mountain's strange shape reminded the Spanish conquistadors of their own deep-hulled galleon ships, so they named the volcano *Galeras*. The "galley" was the carved-out remains of a once-taller volcano that had collapsed and breached the leeward side of the hull. The young cone inside the mile-wide ship was a giant pile of desolate brown rubble with a crater in its center.

Hiking to the volcano was not exceptionally difficult, because the military's access road led all the way to the summit. Local farmers often walked up the road in knee-high rubber boots and thick alpaca sweaters. The campesinos would descend the steep galley wall and walk the half-mile over loose volcanic rock to inspect the rubble-filled crater of their patron volcano. Less frequently, European tourists with heavy backpacks and thick-soled boots made their way up Galeras on the day hike described in their guidebooks.

Not long after the fountain of vapor began to rise from Galeras, the volcano began to rattle. Rocks crumbled from the steep inner walls of the crater, and the military guards felt the police station rumble beneath their feet. The guards sent word to Civil Defense, which passed along a warning to the government in Bogotá, led by the very anxious President Virgilio Barco. Three years earlier, Barco had fiercely campaigned against then-President Belisario Bentacur's management of the Nevado del Ruiz crisis. With the country figuratively crumbling around him, the last thing Barco needed was another Armero.

Although the military guards hadn't seen much change in Galeras during the time that they had been at their post, historically, Galeras was far from being a dormant volcano. In 1831, Jean Baptiste Boussingault gathered the first information on the volcano. The French chemist hiked to the summit of Galeras and found that instead of a crater inside the cone of the volcano, there was a dome of lava 100 feet across—cracked, fractured, and steaming with voluminous amounts of sulfur-laden gas. And as evidence that the volcano had recently erupted, Boussingault noticed that the shrubs and grasses on Galeras's upper flanks were covered with a thin blanket of ash.

In December 1832, Posada Gutiérrez, a commander of the conservative forces from Cartagena, witnessed Galeras erupt and rain ash onto the nearby countryside. The eruption was relatively small, but it likely blew out the lava dome that had filled the volcano's vent the year before. In 1869 and 1870, German scientist and explorer Alphons Stübel studied the 14,000-foot volcano. At the time,

Galeras was experiencing a growth spurt, building a cone like a giant anthill, with small eruptions that blew out rubble, forming a pile 120 feet high from its base on the caldera floor. Stübel noted that the cone was flat on top, made of loose volcanic rock, and had a steaming crater in its center.

By the end of the nineteenth century and the first quarter of the twentieth century, Father José Salvador Restrepo, the director of the Jesuit University in Bogotá, visited Galeras. At that time, the crater was a 30-foot-deep depression, with a muddy clay floor dotted with pits and craters that steamed with sulfur-smelling fumaroles.

Galeras continued to rattle in the early 1900s, and steam poured from the volcano. Then, in 1936, a giant cloud of ash and steam burst from the volcano and shot several miles into the sky. Superhot steam, ash, and rock particles in the eruption column collapsed back into Galeras's amphitheater, creating a small pyroclastic flow that poured over the caldera rim and raced for 2 miles down the northern side of the volcano. The most violent eruption of Galeras ever witnessed was over in just minutes; there were no dead or injured, but the hot pyroclastic flow glowed bright in the night like a thin ribbon of frosting dripping down the side of the volcano. It was a splendid sight never forgotten by the locals of Nariño.

Since the 1936 eruption, Galeras had been quiet, so in 1988, when the mountain once again became restless, the people of Pasto, who lived just 5 miles from the volcano, recalled stories told by their grandparents and great-grandparents. The *Pastusos*, whose potato farms made a patchwork of Galeras's flanks, still held many beliefs of the indigenous population of Nariño. Their native ancestors had named the volcano *Urcunina*—the Fire Mountain—and when billowing steam would rise above Pasto and paint a brilliant picture against the setting sun, elderly *Pastusos* would sit back on their porches and view the beautiful sight. "Urcunina is breathing," they would say. "The volcano is healthy. It is good."

The government in Bogotá, however, did not share this conviction. Around the world, the science of volcanology had changed greatly since the 1985 tragedy of Nevado del Ruiz. There were now a dozen scientists in Colombia who were learning to work with active volcanoes, and the worried President Barco immediately contacted the government-run volcano observatory in Manizales and demanded that someone get to Pasto and fix the problem of Galeras.

Fernando Muñoz received the call. The young geophysics professor had gone to work for the National Institute of Geology and Mines after the eruption of Nevado del Ruiz and was now second in command at the Manizales observatory. "I had just gotten married, and I had no desire to go to Pasto and deal with Galeras," says Muñoz. But the director of the observatory insisted that the young seismologist, who was by far the leader at the observatory, pack his things and go to Pasto.

He was not happy. Besides being a newlywed, Muñoz was a city boy, and Pasto was a small country town. Like the Cafetera region, Pasto and its surrounding municipalities had remained mostly free from the horror of Colombia's civil unrest, but separation and a rural reputation came with a price: *Pastusos* were an endless source of jokes throughout Colombia.

Once in Pasto, Muñoz set up shop at the local university. The 29-year-old seismologist was instantly under great pressure from the national and regional governments to accurately forecast an eruption of Galeras and to figure out what areas of the region could be harmed by the volcano. For several days, Muñoz carefully studied the seismic data. He had been briefly trained by some of the best seismologists in the world, but he knew that all volcanoes had different personalities. Galeras was trying to tell him something. The signals appeared to grow more numerous each day, but Muñoz could only speculate what they meant, and with only one seismometer, it was impossible to determine where they came from. It would take at least three instruments to pinpoint the source of the activity.

When he had been through all of the seismic records, Muñoz and a group of colleagues from Manizales (including Fernando Gil) visited Galeras. They hiked up the gradual slope of the volcano's cone to the crater and then stood on the steep crater rim. The heat was intense, and the scientists tried to stand back from the rim and at the same time look down into the mouth of the volcano. Thirty feet below, the crater of Galeras was a giant glowing cauldron of boiling earth. The entire throat of the volcano glowed in an incandescent orange and appeared to move in rhythmic heaves.

To Fernando Gil and Jaime Romero, who had come to help Muñoz interpret the seismic patterns from Galeras, it was an incredible sight. Gil took a long, deep breath. The scorching glow seemed like the very heart of the volcano, and Gil's pulse raced. "It was nighttime and we were on the rim. It was red—incandescent. We are religious people, and when we saw this, we kneeled down and prayed," he says. Gil had not felt the same sensation since the night he witnessed the eruption of Nevado del Ruiz. Once again, he felt the hand of God reaching into his soul.

Fernando Muñoz, on the other hand, was completely shocked by the fearsome power of the living, breathing volcano. He took a step back from the crater rim.

"I had no idea what the glowing crater meant," he says. Was it molten magma or just superheated rock? Did it mean that the mountain was getting ready to explode? "I had never seen anything like it, but I was pretty sure it was dangerous."

The scientists returned to Pasto in the evening. The first call Muñoz made was to the general director of the National Institute of Geology and Mines in Bogotá. Muñoz listed the things that he had seen happening on the volcano: The seismographs showed increasing activity, as did the fumaroles, and the rocks inside the crater were a fiery red hot.

"I called the director and I told him that I thought the situation was serious," Muñoz recalls. "I told him I didn't know what it meant and that we needed some help."

Muñoz knew that there wasn't any money to spend on monitoring equipment for Galeras, but the disaster of Armero still resonated throughout Colombia, and the director had to answer directly to the president of the republic. He didn't want to take any chances.

"He asked me, 'Who do you want to call for help?' and I told him I wanted to call Dave Harlow and Norman Banks from the

U.S. Geological Survey. I thought they would be the best people to help us."

Muñoz knew from his experience with Nevado del Ruiz how the complicated protocol had to work in order to get scientific help from the United States. First, the director of Colombian's National Institute of Geology and Mines made a formal request for help to the Colombian State Department. The State Department would then make a formal request to the U.S. Embassy, which would then contact the U.S. Office of Foreign Disaster Assistance, which would in turn contact the director of the U.S. Geological Survey. Only then would scientists on the front lines—like Dave Harlow and Norm Banks—be told to pack their bags. If the request for help wasn't punctuated by a catastrophic disaster and thousands of dead bodies, it usually took months. Fortunately for Muñoz, Nevado del Ruiz had been a turning point not only for Colombian scientists but for the U.S Geological Survey too.

At the time he got the call to bring the Volcano Crisis Assistance Team to Pasto, Norm Banks was in the middle of organizing a volcanic hazards workshop. It was originally meant to take place in Chile, but with money from the United Nations Disaster Relief Organization already in place, Banks decided to run the workshop from Pasto and help the Colombians with the Galeras crisis at the same time. While Banks got ready in Vancouver, Washington, Dave Harlow prepared to come to Pasto from the U.S. Geological Survey headquarters in California.

In the weeks that passed, while Muñoz waited for help from the United States, he repeatedly asked if he could leave Pasto and go home to Manizales to be with his wife. His bosses wouldn't hear of it. Muñoz was the official scientist in charge of Galeras. The entire government, all the way up to President Barco, was looking to him for answers. Unfortunately, Muñoz had no idea what to do, and the stress was making him physically sick. Should he recommend that they call for evacuations? Should he tell people that the mountain was getting ready to blow? Muñoz felt he had to do something, so he gathered his fellow scientists together one evening at the university.

"I told my *compañeros* that I was going to send a letter to the president of the republic and tell him that the situation with Galeras is very serious," says Muñoz. "I was going to recommend that people shouldn't be allowed to go near Galeras's crater, and no one should live closer than three kilometers from the crater." He quickly typed and printed a letter to President Barco, confident that he was doing the right thing as he put the letter into the fax machine and dialed the presidential palace in Bogotá. Two hours later Muñoz got a call. It was the director of the National Institute of Geology and Mines, yelling into the phone. Muñoz remembers: "He said, 'What the *hell* is going on? Why did you write that letter? You don't go directly to the president for *any* reason. Are you crazy? This is very serious, Muñoz!' "

Before Muñoz could respond, the director slammed the phone down. "I called my wife in Manizales," Muñoz says. "I told her, 'I think I'm out of a job.' "

Muñoz tiptoed around Pasto for the following week trying to avoid the wrath of the director of the National Institute of Geology and Mines and the president. Finally, Dave Harlow arrived in Pasto, and Muñoz breathed a sigh of relief. "I was sure that Dave Harlow would back me up and get me out of trouble," Muñoz says. Harlow came to Pasto with a team of four scientists and several trunks full of instruments. Muñoz gave the *gringo* a big hug. He hadn't seen Harlow since the American had been at *Piso Once* in 1985, when he had taught Muñoz how to identify a volcanic signature on a seismograph. Harlow was a world-renowned expert. Muñoz hoped the famous scientist could figure out what Galeras was doing and get him out of the mess he was in at the same time.

"I told Dave I was in big trouble," Muñoz says. "I told him I sent a letter to the president of the republic saying that Galeras is likely to erupt, and they were saying that I was crazy. I asked him to do me a favor, to take a look at the seismic records and tell me what he thought."

Harlow and Muñoz spent the whole afternoon looking at two months' worth of seismographs. There was only data from one seismometer, so it was impossible to tell where the quakes and tremors were coming from, but they did look at each signal and analyze it. Some signals looked like small earthquakes—jolts on the seismograph that quickly dissipated; others looked like volcanic tremors—long, thin corkscrews. They wrote down how many quakes happened each day, how strong they were, and how long they lasted.

Finally, Harlow looked up at Muñoz, who had been alternately peering over his shoulder and pacing across the classroom floor. Harlow put a hand on the young Colombian's shoulder. "He told me, 'Fernando, I don't think you should have sent the letter,' " says Muñoz.

Muñoz was speechless. He put his head in his hands for a few seconds, then stared hard at Harlow. "I told him, 'Listen, you're coming with me *right now,* and we're going to the mountain.' It was about six in the evening and we were on the rim of the crater and there were a lot of clouds. We were there for about fifteen minutes, and the clouds cleared and he could see into the glowing crater.

"Dave's eyes got huge and he was backing away. He said, 'Okay, Fernando . . . great . . . I've seen enough. . . . Let's get the hell out of here.' He was nearly running down the cone, so I yelled at him. 'Okay, Dave, tell me what you think.' He told me I was right to send the letter."

Two weeks later, Galeras erupted. It was just an explosion of ash, but it was enough to get Fernando Muñoz out of trouble. "It wasn't a big eruption," he says, "but it was a *real* eruption." He received phone calls from both President Barco and his boss commending him on a job well done.

Next to nothing was known about Volcán Galeras in the scientific community, and the horror of Nevado del Ruiz was still fresh on the minds of the world's volcanologists. But as experts arrived in Pasto to help Norm Banks with his workshop, almost all of them came quickly to an agreement. Galeras was no Nevado del Ruiz. By the spring of 1989, it was becoming apparent that small eruptions were the only thing Galeras was likely to produce over the next few thousand years. Volcanologists called these eruptions "throat clearings," explosions of ash and steam that served to pulverize cold rock plugging the volcano's vent. It would take a million such eruptions to produce one Mount Saint Helens blowout.

The main and most important difference between Galeras and Nevado del Ruiz was that Galeras was not covered by ice or snow. The deadly tragedy of the mudflows that buried Armero was not the direct result of a pyroclastic flow or a lava flow. In fact, the eruption of Nevado del Ruiz had been relatively small, but the heat that it generated acted as a catalyst that melted millions of tons of glacial ice.

Since Galeras offered no threat of mudflows, the concern shifted to pyroclastic flows. Was it possible for an eruption of burning ash and steam to flow down to Pasto and incinerate the population? Studying the written and geological history of the volcano, the scientists came to the conclusion that there would not be a large pyroclastic flow from Galeras, probably not for thousands of years. But in the unlikely instance that they were wrong, the pyroclastic cloud—heavy enough to follow the path of least resistance—would not head toward Pasto, which lay to the east. Instead, it would pour down the western slope of the volcano, where the caldera of Galeras opens up onto the deeply incised and mostly uninhabited Azufrado Valley.

Fernando Muñoz was relieved to hear the consensus. He was also greatly relieved that the volcano was being wired by the best in the business. Although Galeras didn't seem to be the cataclysmic threat people initially feared, it would need constant monitoring in case there were smaller eruptions that could harm the police on the rim or damage the communication towers. And since the government would not make the recommendation to close Galeras to tourists, there was the constant fear that hikers could be killed during a small blast.

Norm Banks and Dave Harlow were setting up a state-of-the-art observatory, soon christened the Observatorio Vulcanológico de

Pasto, in a small house on the outskirts of Pasto rented by the National Institute of Geology and Mines. In the mornings, Banks, Harlow, and Pete Hall, the American volcanologist from Ecuador, taught classes on volcano seismology and deformation to scientists and students. In the afternoons, Banks's team joined Harlow's guys and helped install monitoring equipment on Galeras. They were working nonstop, and the Colombian scientists were in awe of the operation.

The seismographs had already been moved from the university to the new observatory, and telemetered seismometers were constantly relaying the creeks and moans of Galeras back to Pasto, where a seismologist was always on hand, watching and assessing the seismographs as they churned out data from Galeras. If any signal registered on the seismograph, the monitoring seismologist would immediately radio a second team member who was standing on the summit of Galeras. This person, in turn, would transmit the message by radio to the group working in the crater. The relay system was necessary because of the amphitheater shape of the volcano, which made it impossible for the person in the Pasto observatory to be in direct radio contact with the people working inside the caldera.

"We had very strict rules of safety, and we had meetings every evening about coming to a consensus and laying plans. Someone was always standing by the seismographs, and we always wore hard hats," says Dave Harlow. If there was a signal from Galeras, the group inside the volcano would get out as quickly as possible. The work was incredibly dangerous. The scientists had to get very close to the crater to install equipment, and the volcano could erupt at any time. Rockfalls were a danger, and so were the acid gases. The scientists spent as little time as possible inside the volcano.

The Colombian scientists were impressed with the Americans' safety precautions. All of them wore hard hats, gas masks, goggles, and gloves, and they all carried radios. They didn't seem to feel the need to act tough or macho with the volcano, a reckless trait he'd observed among some other scientists. They were actually quite concerned with all facets of safety and often seemed rightfully mindful of the volcano's enormous power. There were always backup plans in case the radios stopped working and plans outlining what to do in the event of an eruption, and if anyone was injured, there were trunks full of medical supplies. Every morning, before anyone went to work in the field, Norm Banks would hold a state-of-the-volcano meeting and discuss what they had learned the previous day and what concerns they had about the volcano's future.

Fernando Muñoz knew that they had good reason to be overly prepared: The volcano had just erupted two weeks before. It was a small eruption, but anyone near the crater would have been blown to pieces as small as volcanic ash. The scientists were also worried about the possibility of a pyroclastic flow—an absolute death sentence that kills not from the heat but from inhalation of scalding hot ash. On the first breath, a person's lungs react with instant pneumonia and fill with fluid. With the second breath, the fluid and ash mix and create wet cement. By the time the person takes a third breath, thick, hot cement fills the lungs and windpipe, causing the victim to suffocate. There were autopsy pictures of a surgeon opening a victim's trachea with a chisel. And ever since Norm Banks had seen how his colleague David Johnston died at Mount Saint Helens, he always carried three small cans of pressurized air—the emergency canisters used by scuba divers. He planned to bury himself if he could, and wait for the pyroclastic flow to pass. Some of his fellow volcanologists thought Banks was a little overcautious, but he wasn't taking any chances.

After the first week of the workshop, Banks, Harlow, Pete Hall, and the rest of the scientists were completely exhausted. They were dealing with the crisis of Galeras as well as all of the complicated scientific and political factors that come with an active volcano. The mayor of Pasto and the governor of Nariño were officially supportive of the scientists, but when the scientists would hold meetings to address the safety issues for Pasto, neither politician would attend. They did not want to worry their constituents, so they played down the scientists' recommendations for emergency preparations.

Day after day of strenuous work at 14,000 feet was exhausting, and the U.S. scientists were getting very little sleep. Dave Harlow was beginning to think the whole thing was a bad idea. "It was a bit of a nightmare. The truth is that we were all willing to do that kind of work, but Norm was becoming a little aggravated, and when we would joke and point it out, Norm's reaction was to take it very personally."

By the time the workshop was over in May 1989, bad feelings solidified between Norm Banks and his colleagues. Banks never had been able to play politics within the U.S. Geological Survey, and the result was that he would soon be forced to resign from the very program he had worked so hard to create.

At the same time that the foreign and Colombian scientists were consumed with the volcano, the people of Nariño province, who had lived in harmony under the shadow of Galeras for thousands of years, were having their lives turned upside down. In March 1989, with no explanation to the locals, scientists and politicians began declaring color-coded volcano alerts that were broadcast over local radio and television stations. First, the scientists, led by Fernando Muñoz, suggested that the authorities call a white warning. But a white warning had no meaning at all. It just seemed more benign than calling a yellow warning—something the uncertain scientists were afraid to do. The same day, a risk committee of Nariño politicians and civic leaders declared a yellow warning anyway, which was broadcast over Pasto radio. According to the color-alert scheme, it meant that the volcano was showing signs of activity, but the risk committee failed to explain this to the media, who simply broadcast a yellow alert, without any kind of explanation. The following day, the government in Bogotá, fearful of another Armero and without consulting officials in Pasto, declared an orange warning and planned to call for evacuations. Orange signified impending disaster, but again, the *Pastusos* had no idea what any of it meant and there were no directions on what to do. The national media

picked up on the orange warning with gusto, and were soon reporting that Galeras was ready to incinerate the entire population of
Nariño. Making matters even worse, the scientists at the volcano
observatory created a hazard map with concentric rings around
Galeras like a bull's-eye that placed the entire western half of Pasto
in a high-hazard zone.

Immediately, national banks quit giving loans to Pasto residents.
The farmers, who survived season to season by taking out short-
term loans, were unable to buy seed, fertilizer, feed, and equipment.
New construction ceased completely—why would investors back
builders in a town that was sure to be wiped off the map at a
moment's notice? Tourism, once a stable addition to the economy of
the quiet region, totally dried up. Companies from outside Nariño
refused to send their trucks into the area to deliver goods, and market shelves were soon empty. *Pastusos* who could, sent their children to live with relatives, and property values plummeted as fearful
families living close to the volcano packed up and moved from
homes they were unable to sell. The volcano that had always been
their friend had turned against them overnight.

Omar Darío Cardona hoped he could do something to fix the
economic crisis that was becoming a disaster on its own. Cardona
was a bright young civil engineer from Manizales who had been
working for the government since 1987 to help deal with his country's seemingly endless natural disasters. Floods, earthquakes, and
disease had plagued Colombia during his two years as second in
command of the National System for Risk Mitigation and Disaster
Preparedness—the agency he helped spearhead after the devastation
of Armero had changed his life. Now, another Colombian volcano
was waking up, and Cardona was determined to make sure it didn't
end in catastrophe.

As a young boy, Omar Darío grew up in a comfortable house at
the north end of Manizales. His bedroom window looked out onto
Nevado del Ruiz and its majestic, snow-covered summit. Omar
Darío knew that the giant mountain was a volcano, but no one
in Manizales, including his parents, ever spoke of it as a source of
danger.

One day, when he was a teenager, Omar Darío woke up frightened from a terrible nightmare. In his dream, the great volcano erupted, and a giant flood raced toward Manizales. It was the only dream he had ever remembered, and it was so vivid that he almost believed it was real. But volcanoes, he had always been told, produced giant explosions of molten rock and ash, not floods of water. The boy became obsessed with the volcano and never could erase the nightmare from his consciousness.

In the summer of 1985, Cardona had the opportunity to go to Europe to earn a graduate degree in earthquake engineering. He was walking the streets of Barcelona on November 15 when he passed a newsstand. On the cover of a Spanish newspaper was a black-and-white photo of the buried city of Armero and thousands of the tragedy's mud-encrusted victims. Nevado del Ruiz had erupted, and the result was a giant flood. It was so similar to his nightmare, he could hardly believe it was true.

When Cardona came back to Manizales in December 1985, the tragedy of Armero continued to haunt him, and he spent a great deal of time investigating what had happened leading up to the fateful day. "I was very hard on the government, and I blamed it for the disaster," Cardona recalls. The young engineer went on a crusade and became a relentless antagonist to the government, loudly and vehemently criticizing how the entire operation was handled.

"This is the fault not of a volcano but of an incompetent government!" he would yell at seminars in university halls and public meetings, waving his arms about madly. "The government completely missed the mark," he would say, pausing for effect between sentences. "The scientists missed the mark. No one was taking into consideration the people *or their culture*."

After several months of tirades and blistering accusations that often ended up in the newspaper, Cardona was called to Bogotá by the newly elected president, Virgilio Barco, for a meeting. "Well, Señor Cardona," said the president, "since you seem to be so adamant against the government's approach in dealing with the volcano, why don't you tell us what we should do?"

It was an invitation for an hour-long monologue by Cardona.

The president, fearful of another disaster like Nevado del Ruiz, was adequately impressed, and by 1987, the young engineer had helped organize the National System for Risk Mitigation and Disaster Preparedness.

In 1989, by the time he was sent to Pasto to deal with the threat of another volcano, Cardona had already seen many disasters in his short tenure with the agency. Floods plagued the high valleys. Hurricanes dumped rain and destroyed millions of dollars of property on the Pacific coast. Enormous landslides buried entire villages in the Andes. Cardona even found himself dealing with a cholera epidemic in the tropics. Each new disaster was an intense learning experience, and with each event, Cardona felt that the government, the scientists, and the media were all working at cross-purposes— with the people caught helplessly in the middle.

Things were no different in Pasto. Initially, the mayor of the city was very supportive of the scientists, but the support soon waned when the mayor's constituents began to complain loudly about what they considered to be fear-mongering by the scientists. They had lived their entire lives with Galeras, and they were sure that the volcano could do no harm.

Cardona felt like he was on a constant mission of damage control. He was the fragile link between the scientists, the local politicians, the media, and the national government. "I made numerous visits to the towns surrounding Galeras. The mayors were my friends also. They used the crisis to ask for resources, to get better services and a road, but they never believed in the hazard. In four hundred and fifty years there had not been a single victim of the volcano."

In May 1989, Cardona flew to Bogotá just in time to dissuade the national emergency committee not to evacuate Pasto and the areas that surrounded the volcano, explaining to the worried politicians that a massive eruption that would harm Pasto was statistically unlikely. The committee agreed to hold off the order to evacuate.

It was a close call. Cardona could only imagine the mayhem that a wholesale evacuation would cause. Pasto was built into a valley surrounded on all sides by mountains, and there was only one narrow road leading out of the city. There were no temporary shelters set up to house those who couldn't stay with relatives outside of the hazard zone. The people of Pasto were not wealthy. Many didn't own vehicles. For the farmers, leaving their land unattended would be a disaster in itself.

Cardona was shaken when he left the committee meeting. He had only dealt with Galeras for a month, and he was already exhausted. Unlike hurricanes, floods, and earthquakes, which had a beginning and an end to their destruction, the life span of a volcano crisis was completely different, and Cardona could see no end in sight.

By July 1990, the new governor of Nariño, promising to deliver Pasto from the economic ruin it had suffered for nearly a year, forbade information concerning the activity of Galeras from being released to the national media. Press releases issued by the observatory were never distributed. The news correspondents reacted to the embargo with new fervor, openly questioning the official silence and generating speculation among *Pastusos* that the government had something to hide. The locals had long thought that there were other motivations behind the scientific assault on their volcano. It was hard to believe that anyone would actually *work* as a geologist, and as soon as the scientists showed up in Pasto, rumors abounded that there was actually gold in the volcano, and the government was trying to hide it from the people.

The scientists, who came rotating into the observatory from their jobs in Manizales, Bogotá, and Popayán, had good intentions, but they had no idea how to talk to the journalists who were constantly pestering them with what seemed like the obvious question: *When will Volcán Galeras erupt?*

The volcanologists from the observatory (who were now under the direction of an increasingly frustrated *Bogotano* named Jaime Romero) handed the journalists a technical scientific report. The local media, completely unable to decipher the jargon, decided to publish the report in its entirety in the local newspaper. No one in

Pasto had the slightest inkling what it all meant, and it deepened the rift between the scientists and the *Pastusos*. Romero and his fellow scientists began to feel hostility from the locals: "The scientists were not accepted by the people of Nariño, and when they saw us, they turned up their noses. They would say, 'Why do these people come here to see the volcano?' They looked at us like we were strange." Nevertheless, the locals were becoming worried, and Romero received daily phone calls from *Pastusos: Please call and tell us when the volcano is going to erupt so we can leave.*

But Galeras failed to produce the promised deadly eruption. The authorities and the scientists scratched their heads and began to doubt that the volcano would ever actually hurt anyone at all. The small eruptions of ash and gas that had occurred looked just like what was described in historic reports and what was known through oral tradition.

Looking to point an accusatory finger for the economic misery that had befallen Pasto, the chamber of commerce and the local government held Jaime Romero, Marta Calvache, Fernando Muñoz, and the other young scientists at the observatory responsible for the mayhem. They were accused of managing information poorly and blasted for using inappropriate educational materials, such as slide shows of a wildly exploding Mount Saint Helens that terrified *Pastusos*. "We didn't know how to handle the communication aspects, and I felt responsible," says Muñoz. "I didn't have the experience to do the communications. Today I think that there was too much emphasis on the work that we were doing. The work that we were doing was only about science." To make matters worse, the Nariño governor openly ridiculed the scientists. "Marta was going to speak in public with the governor and she was wearing a shirt with a picture of a volcano on it, and the governor said, 'Look, she's crazy, she even wears a shirt with a volcano on it,'" says Muñoz. The locals picked up on the politician's spin of the crisis and angrily blamed the scientists for the invented curse of the volcano.

By late 1991, a new governor was elected in the department of Nariño, and a new mayor in the city of Pasto. Both politicians aggressively campaigned against the incumbents by stressing the administration's poor management of the volcanic crisis and the economic recession that followed.

Meanwhile, Galeras, having remained relatively quiet for nearly a year, began to rattle again in November 1991. Viscous magma was rising in the throat of the volcano and building a dome in the crater. It was something never before seen on Galeras in recent times, so before he left office, the retiring governor called a meeting of U.S. and Colombian scientists to analyze the volcano and report what they had found.

Marta Calvache, who was next in line to take over running the observatory, was already in Pasto for the meeting. She had spent the previous three years between Colombia and the United States, where she was finishing a master's thesis under Stanley Williams at Louisiana State University. Calvache had met Williams after the eruption of Nevado del Ruiz. She had enjoyed living in Baton Rouge and had fallen in love with the Cajun food but missed the cool climate of her Andean homeland. "Originally, I planned to study Nevado del Ruiz for my thesis, but once Galeras began to wake up, I changed my mind," Calvache says. Now back in Pasto, she was working like a detective to decipher the million-year geological history of Galeras by gathering clues in the layers of rocks that built the volcano.

Unlike many of her big-city colleagues, Calvache *wanted* to work in Pasto. The diminutive scientist with the heart-shaped face and giant blue eyes was born in Consacá, a small farming village on the western side of Galeras. When she was just 2 years old, her father passed away, and her mother was left to raise Marta and her two older brothers and sister. The work on the farm was extremely difficult for the young widow, but in the early 1960s, the Nariño countryside was a peaceful place to raise children.

As a young girl, what Marta knew about the tall mountain rising to the east came from local legend: *From time to time Galeras will wake up and exhale white clouds and beautiful fountains of fire, then he will go back to sleep.* From that small wooden house where she lived, the volcano looked like a giant slide, with the open caldera of Galeras spilling into the steep Azufrado Valley. Galeras never had so much as a hiccup while she was growing up, though Nariño villages near Galeras were commonly rattled by earthquakes.

Calvache was the obvious choice to head the observatory in Pasto. She was gracious and intelligent, and she made sure that the observatory staff provided any assistance needed to the foreign scientists who came to work in Colombia. Despite her diminutive size, she was tougher and stronger than most of the men. Ever since the days of Nevado del Ruiz, Calvache had taken to mountains as if they were mere foothills, often leaving foreign scientists scrambling to keep up with her in the field.

Returning to Colombia after being in the United States was always somewhat of a shock to Calvache. In Pasto, the observatory had money for only one telephone line, and it had to be shared with the fax machine. The locals would steal seismometers for parts. Worse still, the mayor and the governor had no respect for the scientists. It was all a little difficult to take. She knew what a complete headache the observatory had been for Jaime Romero, but Calvache was drawn to Pasto and to Galeras, and she was definitely not afraid of a challenge.

Joining Calvache and several others for the Galeras meeting was Bernard Chouet. The Swiss-born seismologist from the U.S. Geological Survey had come to Manizales six months after the Armero tragedy to try to unravel the confusing collection of theories about what Nevado del Ruiz had in store for the region.

At the time, Chouet had worked with Fernando Gil on the details of a new theory that he had been developing—a discovery that could be used to predict whether a volcano was getting ready to blow. Together, Chouet and Gil mulled over seismographs collected before the November 13, 1985, eruption. The carbon-coated records were covered with small signals that had the shape of long metal screws. These were the marks that Gil and Fernando Muñoz had thought were just malfunctions of the seismometer.

Chouet put a thick finger on one of the marks. "These signals," he said, "are called long-period events."

For years, seismologists had been seeing the strange, screw-shaped signals in seismograph data from active volcanoes. They appeared before the 1980 eruption at Mount Saint Helens. They swarmed the seismographs on a Japanese volcano named Asama in 1983 and El Chichón in Mexico in 1982, just before they erupted. But the seismologists who looked at the data were looking for the short, spastic signals generated by earthquakes—a sign that magma was rising to the surface—and they had no idea what the long, thin, spiral etchings meant.

Up to that point, seismologists seemed primarily concerned with cataloguing and naming the different types of shapes that the long signals made. Bernard Chouet thought it a ridiculous way of handling a scientific problem: Rather than trying to determine what was obviously the result of a physical phenomenon inside an active volcano, what the discipline was doing "was more like cataloguing flowers or butterflies," says Chouet.

Chouet believed he knew what the signals were saying. Inside the volcano, in fractures in the rocks, boiling water turned to steam. And the steam, under great pressure and unable to escape, resonated brutally in the fractures, creating a high-frequency song like a boiling teakettle whistling an imperceptible pitch. The vibrations were enough to rattle the rocks around them, and the reverberations would show up on a seismograph. What Chouet had discovered, and proven through a series of complex equations, was that when a volcano appears to rest peacefully on the surface, it could actually be under immense—and deadly—pressure. This theory differed markedly from current thought, which was that when a volcano looked quiet on the surface, and the seismographs didn't show earthquakes rupturing beneath the cone, the volcano was going back to sleep.

When Fernando Gil first heard Chouet's theory, he was fascinated. He had never heard anyone offer an explanation for the long, thin signals. They had always been treated as background vibrations or problems with the instrument—and all the while Nevado del Ruiz had been warning them of a disaster.

In the years after Nevado del Ruiz, Bernard Chouet continued to fine-tune his discovery. Within the U.S. Geological Survey, scientists were divided into two unofficial categories: Cone Heads and Meatballs. Cone Heads were the analytical types, the thinkers—comfortable in dark offices under the green glow of computer screens. They published the scientific papers, got the promotions, and were heralded in the community. Meatballs were scruffy, prone to upsetting their bosses, and much more at home climbing mountains or sniffing around for minerals and oil.

Among the USGS scientists, Bernard Chouet was considered a Cone Head of the highest order. By 1989, he had been working for the U.S. Geological Survey at the West Coast headquarters in Menlo Park, California, for five years and had mostly kept to himself—more comfortable with his endless equations than with the Meatballs and their boisterous camaraderie. The Meatballs left him alone for the most part, until one day in December 1989. "I was waiting for the copy machine behind Bob Page, and he told me he had something he wanted me to take a look at," says Chouet.

Page and two other seismologists marched into Chouet's office with a stack of seismic data printouts.

While Chouet mulled over the records for several minutes, Page told him that the data came from a volcano in Alaska called Redoubt. The printouts were full of weird signals, and no one had been able to figure out what they meant.

"I asked him what the volcano looked like on the surface, and he told me it appeared to be quiet," Chouet says. He spent a few more minutes browsing through the records. "Then I said, 'I think that your volcano is getting ready to blow.' " The Meatballs held back their laughter.

The next day, Chouet ducked his head into the seismic laboratory. "I said, 'Hey, how's your volcano? Has it erupted yet?' They told me that they hadn't checked in that day, but if I was so worried about it, I could call up there and ask them."

They gave Chouet the number for the Alaska Volcano Observatory, and he went back to his office. "I called up there, and I said, 'Hello. This is Bernard Chouet calling from Menlo Park.' The person on the other end cut me off and said, 'Listen, I can't talk right now. The volcano is erupting.' "

Chouet put down the phone. He had seen the long-period events in data that had been collected before eruptions on Mount Etna, Nevado del Ruiz, and Mount Saint Helens. But he had always been looking at the data *after* the fact. This was the first time he had actually been able to predict an eruption.

After Redoubt's first eruption, the Alaskan volcano proceeded to have small eruptive spurts, as magma pushed its way up to the volcano's throat. The scientists in Alaska sent daily seismic reports to Chouet, and he watched them carefully. When magma was rising to the surface, the seismic signals were markedly different from those produced during long-period events. They were short, jagged zigzags—records of rocks fracturing as magma broke through them. The patterns were similar to what seismometers recorded when an earthquake fractured rocks below the earth's surface. But by the end of December, the seismic signals coming from Redoubt had changed from volcanic earthquakes back to long-period events. The throat of the volcano was plugged and pressurizing, and Chouet was nearly certain that Redoubt Volcano would soon blow again.

The lava dome had grown so large that there were 40 million cubic yards of rock hanging dangerously over the west side of the volcano in the direction of an oil production platform, and Redoubt was covered by glaciers. As with Nevado del Ruiz, the combination of ice and an eruption could be devastating, but shutting down oil production to wait for an explosion that might never come would cost millions. Tom Miller, the scientist in charge at the Alaska Volcano Observatory, decided to go with Chouet's recommendation, and on January 2, 1991, with long-period signals swarming the seismographs, he called for an evacuation of the oil terminal. Production was shut down, and the workers were removed from the site.

Two hours later, Redoubt Volcano exploded in a gigantic eruption, melting its colossal glaciers and creating an enormous mudflow that reached all the way to the oil terminal. The damage to the terminal was costly, but there were no oil spills, and no deaths—not even a single injury. The oil company was elated with the scientists' predictions. "The next day, Tom Miller called me and said, 'These guys think we walk on water,'" Chouet says.

After Redoubt, Chouet published several papers on his work in highly scientific journals and presented his data at scientific meetings. The work was still very new, and despite his success at Redoubt, few scientists in the world were even aware of what Chouet was up to.

When he arrived in Pasto in November 1991 to meet with Fernando Gil, Marta Calvache, and several others, the seismicity at Galeras was at an all-time high. Along with Gil, Chouet hiked to Galeras's crater and stood on its rim. A clot of thick lava with the consistency of wet cement had risen to the surface and formed a solid dome in Galeras's crater, plugging the volcano's vent like a cork. A long, thin crack ran the length of the lava dome through its center. The scientists surveyed Galeras for the entire morning. Every hour or so, pressure would build up beneath the dome, and the crack would open slightly and let out a puff of ash and steam.

Chouet's assessment was that for now, the volcano is safe. But he was sure it wouldn't stay that way. He reported to the authorities that soon the crack would seal, and Galeras would appear quiet on the surface. But underneath, pressure would be building and the seismic signals would turn into long-period events. Then, without warning, there would be a small eruption that would blow apart the dome. It would be deadly for anyone nearby, but not pose much of a threat beyond the immediate area of the mountaintop. Chouet recommended moving the communication towers and the police hut to a safer location and lighting the runway of Pasto's airport to prepare for a possible emergency. The city of Pasto, he said, in agreement with the reigning scientific consensus, was not in any immediate danger.

Chouet's report also called for restricting tourism near the crater and for better security for the volcanic monitoring equipment, some of which, powered by expensive batteries, had been damaged or stolen. The report was filed with the local and national government and the observatory, and Chouet left for home.

Months later, tourism hadn't been restricted. Jaime Romero, Marta Calvache, and the other observatory scientists believed in Chouet's recommendations, but they were simply impossible to carry out. Who was going to restrict tourism? It costs *money* to hire guards. And move the radio towers? *Money.* And lighting the airport? The Colombians appreciated the suggestions, but sometimes the Americans just didn't get it.

In mid-1992, Galeras began to appear quiet on the surface, and the seismicity had calmed down. By July, the crack in the dome had sealed shut, and strange signals appeared in the seismographs. Later, the Colombians named the long-period events *tornillos,* the Spanish word for screw. There were nine *tornillos* over a period of five days.

Then, on July 16, Galeras cleared its throat and pulverized the solid cork that had been plugging its vent. Gigantic boulders were heaved into the sky and a cloud of ash went 3 miles into the atmosphere. Atop Galeras's rim, several of the communications towers were destroyed.

It was the exact scenario forecast in Bernard Chouet's report eight months earlier.

CHAPTER 8: DISSECTING GALERAS

PASTO, COLOMBIA:
JANUARY 6–13, 1993

IT WAS JANUARY 6, 1993, THE LAST DAY of the Carnival de Negros y Blancos, celebration of a day of rest for African slaves mandated by the king of Spain in 1607. Marta Calvache looked out the second-story office window of the tan stucco observatory building. It seemed like the whole city of Pasto had emptied out onto the street below. A melee of costumed revelers danced in circles; vibrant tigers and fearsome ghouls sang along to music blaring from giant speakers set in the doorways of closed storefronts.

In just five days, Calvache—and Pasto—would host the biggest international scientific gathering in Colombia's history. The International Workshop for Volcán Galeras was the brainchild of Stanley Williams, Calvache's graduate adviser whom she had followed when he left Louisiana State University for Arizona State University. Three days of scientific talks and presentations, a day of field trips, over 100 participants. "We had to prepare everything," Calvache says. "It was very difficult. We were working inside, trying to ignore the people dancing outside."

Grabbing a thick stack of papers, she walked down the narrow staircase and out the front door of the observatory. She hurried into her car, dodging squirt guns and fistfuls of flour thrown about like confetti.

The observatory didn't have its own photocopy machine—it couldn't afford one. And there wasn't a single store open in the entire city of Pasto. In fact, the whole state of Nariño was closed for business. As a last resort, Calvache knew a local businessman with a copier in his house, but he lived clear on the other side of the city. Every time she needed to make copies for the workshop, she had to stop working, get in the car, and make the fifteen-minute trip. Calvache's car squeezed its way through the fray of partygoers and arrived at a small house on the east side of Pasto.

She had only been back in Pasto for a week after passing the oral exams for her doctorate in geology at Arizona State University. Now that she was back, she wished she had a few days to relax—to celebrate with the rest of the city and catch up on her sleep. Yet there she was, stuck either toiling in her small, cramped office or fighting with this copier. *No wonder all the locals think we're strange.*

Hours later, as the sun was setting, she loaded her car with cardboard boxes full of copies and drove back to the observatory. The carnival continued to rage in the streets; *Pastusos* stumbled wildly in front of her car. Bruno Martinelli was at the observatory when she returned. He was sending last-minute faxes to many of the scientists who would arrive in four days, on Sunday, January 10, the day before the workshop would begin. *Thank God for Bruno,* Calvache thought. Martinelli gave Calvache a wink as she carried a box full of papers upstairs to her office.

When Nevado del Ruiz had begun to rumble in 1985, Martinelli, the Swiss seismologist, had come to Colombia carrying a trunk full of seismometers. He had fallen in love with Colombia and with Marta Calvache. Nearly two decades his junior, Calvache had become completely enamored of the big, flirtatious Swiss scientist when they met in *Piso Once* in 1985, and the two had been a couple ever since. But in the eight years they'd been together, they hadn't spent more than a few months at a time in the same country. While Calvache had been back and forth between Colombia and the United States, Martinelli spent his time headquartered in Switzerland and traveling all over the world dealing with volcanic and earthquake disasters.

There was only one way that they could be together permanently, and it meant that Martinelli's career as one of the foremost volcanologists in Switzerland would take a backseat to making a life in Colombia. Calvache had been assured the job as director of the Observatorio Vulcanológico de Pasto, and Martinelli had promised to join her. But by October 1992, when Martinelli secured funding to work with the Colombians in the sleepy southern town, Calvache was on her way out of the country to work on her Ph.D. with Stanley Williams at Arizona State University. It was ironic, she thought. Finally, they were both in the same place at the same time, and they *still* had no time to spend together.

The previous week, though, Calvache and Martinelli had managed to spend one morning together when they went to Galeras to scout for field trips. After three days of talks and meetings in the conference room of a local hotel, the scientists would be offered a field trip on Thursday, January 14. The conventioneers would have several options, but Calvache and Martinelli were sure that many would opt for the tour of the very inside of Galeras, for an actual look into its steaming crater.

As far as volcanoes went, hiking Galeras was a breeze—nothing like the treacherous trip up Nevado del Ruiz. But there was one very steep slope that had to be descended and ascended with the help of a fixed rope, and Calvache and Martinelli wanted to make sure the way was as safe as possible. The last thing they wanted was someone getting hurt during the workshop.

The weather on top of Galeras changed by the minute, and it was easy to get caught in the crater when the visibility went to zero. So Calvache and Martinelli marked boulders along the way with orange spray paint. On their return trip back up the rope to the summit, they negotiated around several large boulders jutting precariously from the steep cliff. Martinelli was worried. "It looked dangerous, like the rocks could easily dislodge from the slope and cause a rockslide," says Martinelli.

At the summit, Martinelli hiked to the military outpost. There, he remembers, "I asked them to dynamite the slope before the field trip the following week. The sergeant told me, 'No way. We can't do that. If we set off dynamite, it'll make Galeras erupt.' I told them that we were going to have many people in the crater, and that the rocks were dangerous." But the sergeant wouldn't relent.

Martinelli was flabbergasted. He had been in Pasto only three months, but it hadn't taken him long to realize that the sentiment was the same throughout the region.

"I came to Colombia to help with the hazard of Galeras," says Martinelli. "But I realize now that there are more fundamental social and political issues that needed to be the priority here." How do you make people care about the distant possibility of an erupting volcano when food is scarce, the economy is based mainly on an illegal cocaine trade, and thousands of people a year are murdered by their own countrymen?

Stanley Williams had been working in Colombia since the aftermath of Armero to bring Colombian volcanoes onto the radar screen of the world's volcanologists. With funding from the National Science Foundation, he had been able to install expensive gas monitoring equipment on Galeras and Nevado del Ruiz. And although he was sometimes difficult to work with, many of the Colombian scientists appreciated his seemingly genuine interest in helping them avoid another volcanic tragedy and his efforts to bring them into the world of mainstream scientific research.

Williams's arrival at the conference had been delayed three days while he waited in Bogotá for safe passage to Pasto. There were only two ways to get to the small southern city: by plane and by car. Air travel was hit-and-miss because the small airport outside of Pasto, equipped with neither radar nor lights, was stuck in the middle of the Cordillera Central, bordered by mountains and sheer cliffs, and often completely socked in by clouds. The only other way was along the highway that ran from Bogotá to Cali to Pasto, but because of the large numbers of bandits and guerrillas who were known to prowl the route, it was considered the most dangerous road in the entire country—definitely no place for *gringo* scientists.

Calvache turned on her computer, and as she waited for the slow machine to boot up, she remembered that she also had to change the field trip day on the workshop schedule. The power would be running at 50 percent on Thursday—a scheduled brownout to conserve energy because of economic problems in Pasto. The hotel owner had called to warn her, knowing that the scientists would need extra power to run the slide projector and light the meeting rooms.

"Could you change your schedule and go on your field trips on Thursday instead of Wednesday?" he had asked.

"Of course," Calvache had said. What could one day matter?

Around the world, earthquakes, volcanoes, and hurricanes had turned the 1980s into an especially catastrophic decade. Hurricane Hugo pummeled the U.S. southern coast and the Caribbean. Floods devastated parts of India and China. Regions of Armenia, Mexico City, and San Francisco were reduced to rubble by massive quakes. More than 700,000 people had died and over $160 billion was spent to clean up what seemed like one disaster after another. And in Colombia, in the city of Armero, the official number of dead came in at just over 23,000, ranking the 1985 tragedy as the second deadliest volcanic disaster of the century.

As the 1980s came to a close, the United Nations proclaimed the 1990s the Decade for Natural Disaster Reduction, calling upon international scientists to look into what could be done to ease future losses.

With Galeras having minor stints of explosive behavior since 1989, Stanley Williams suggested using the United Nations' call for disaster reduction to help secure money to study the sputtering volcano. By late 1992, Galeras, along with fifteen other volcanoes around the world, had been nominated as a "decade volcano" and singled out for intense study as part of the Decade for Natural Disaster Reduction.

While Calvache studied for her oral exams at Arizona State University, Williams had invited the biggest names in the volcano business from all over the world. Excited by the prospect of a look at the active Colombian volcano, many said they would come. Williams was especially interested in gathering people like himself, whose specialty was the chemistry of volcanic gases. He touted the ease of collecting gas samples at Galeras to encourage Fraser Goff, a chemist from Los Alamos National Laboratory, to attend the workshop.

"Galeras," Williams told Goff several months before the meeting, "is a piece of cake."

Sunday, January 10, was beautiful and clear, and the airport, which sat on a ridge 25 miles north of Pasto, was finally open. Andy Adams, a big, good-natured technician—part of the team from Los Alamos—and his boss, Fraser Goff, hailed a taxi at the airport entrance. The route to Pasto was narrow and winding, and the broad sloping cone of Galeras occasionally jutted into view. Its brown slopes were covered in green grasses that gave the mountain a soft, peaceful look, and channels forged by rainwater had carved wrinkles in the volcano's sides, recording the passage of time in its million-year-old edifice.

Pasto is a quaint city nestled in the narrow Atriz Valley 9,000 feet above sea level and surrounded by jagged green peaks and the lumbering 14,000-foot volcano. Pretty colonial churches and white stucco shops line the narrow streets, where local artisans sell decorative pottery on sidewalks etched with the signatures of past earthquakes. Pasto's 300,000 residents take advantage of miles and miles of rich soil draped over crested mountains; the economy is based almost completely on agriculture.

In 1539, less than a century after the Andean highland natives had been overrun by Incas, Lorenzo de Aldana came north from Quito to establish Pasto as one of the first Spanish colonies in Colombia. By 1993, most of the early colonial architecture had succumbed to a series of strong earthquakes, and locals were well aware that treacherous *terremotos* could come in an instant. The earth would quake fiercely beneath them and turn their homes to rubble. But of the massive volcano that loomed above them, they were forever unafraid. Galeras belongs to the people, they would say. We are his family. He is our protector and our friend.

The taxi carrying Andy Adams and Fraser Goff pulled up in front of Hotel Cuellar, a moderately upscale establishment near the city center where the workshop would be held. They unloaded their equipment and climbed the stairs to the hotel foyer. The lobby was already full of North American scientists in bright Gore-Tex jackets and expensive field boots, their fancy backpacks adorned with sleek logos. Other scientists dressed in thick wool sweaters, worn work boots, and flannel shirts; still others wore casual business attire.

There were scientists representing more than a dozen scientific disciplines at the gathering. There were seismologists, who listened to a volcano's creaks and moans to interpret signs of unrest; deformation experts, who watched a volcano's bulging sides to see if magma was rising to the surface; geologists, who mapped the remnants of past eruptions in order to predict the future; chemists, who sampled water and steam to see if the volcano's circulatory system was changing. There were many other types of scientists too—some who looked at craters on the moon or looked at the Earth from satellites—who were beginning to find that they could add something to the wealth of volcano research.

The scientists retired early that night, and the next morning—the first day of the conference—they gathered in a large meeting room on the first floor of Hotel Cuellar. It was a reunion of sorts for many of the scientists. Nestor García had come from Manizales, where he taught at the university. Calvache hadn't worked with García since the two of them were climbing up to Nevado del Ruiz together in 1985, and Calvache was happy he'd made it to the conference. Fernando Muñoz drove to Pasto from Manizales with Christopher Sanders, one of his professors from Arizona State University, where Muñoz was finishing a master's degree in geophysics. Maria Luisa Monsalve, Marta Calvache's best friend from college, came, as did Fernando Gil, who had become the foremost seismologist in Colombia. But for most of the workshop participants, this would be their first view both of Galeras and of the new Pasto observatory, which had recently relocated to an office building on the corner of Calle 18 and Carrera 31 in the center of town.

Andy Adams headed into the cold conference room. The presentations were to begin at 9 A.M., but the first two speakers—the mayor of Pasto and the governor of the state of Nariño—didn't show, which was to be expected. (Marta Calvache had invited them out of diplomacy, but relationships between the scientists and the local politicians were impossibly bad.) So wide was the chasm that the mayor was still refusing to even admit that Pasto had a volcano in its midst.

The meetings got under way, with talks given in either English or Spanish, sometimes both. As is typical during scientific meetings, audience members in the dimly lit room periodically dozed. When the talk was in Spanish, Andy Adams would try to glean what he could from the overheads and slides. Eventually, though, he gave up trying to comprehend the Spanish talks and went out to the lobby for a cigarette.

At the morning coffee break, the hotel lobby became raucous as excited conventioneers swapped field stories and scientific ideas in six different languages. Adams and Fraser Goff spent part of the break with Igor Menyailov, a skeletally thin Russian with whom they had worked on several occasions. "Igor was a really crusty old fart," says Goff. "He'd worked out there in Kamchatka and was one tough motherfucker. He knew it and everyone else knew it, too."

After the break, Adams walked back into the crowded meeting room where Setsuya Nakada and Tad Ui, two Japanese geologists, were giving a presentation on Mount Unzen, a temperamental Japanese volcano that had recently incinerated forty people in a pyroclastic flow. Nearly all the people who died were journalists; all forty had been inside the evacuated zone, after making the tragic decision to follow two French volcanologists who were trying to get a better view of Unzen erupting. In his presentation, Nakada showed a slick video produced by the Japanese Geological Survey. The story of Unzen was gruesome, but Adams couldn't help but marvel at the amount of money that the Japanese had to spend on volcanology.

By Tuesday, the second day of the conference, the talks were getting somewhat tedious to Adams. Many were in Spanish, and he was growing weary of sitting on the hard chairs in the meeting room and trying to figure out what was being discussed. By midday, Gudmundur Sigvaldason, a giant Icelandic volcanologist, gave a presentation on smoothing the waters between governments and scientists when dealing with possible volcanic disasters. This was a nice change of pace from the technical science talks. The story he told of Soufrière Volcano on the island of Guadeloupe in the French West Indies transfixed not just Adams, but the entire audience.

When the 5,000-foot stratovolcano on the south side of Guadeloupe began to show signs of activity in 1976, two groups of French scientists were sent to the modest volcano observatory on the island. One group was led by Haroun Tazieff, the phenomenally difficult official who in 1985 had commandeered the helicopter at Nevado del Ruiz. The other was headed by Robert Brousse, another prominent French volcanologist and Tazieff's scientific archrival. Tazieff went to the volcano, looked around, and made a proclamation that it was harmless. Brousse, however, was convinced that the volcano would surely blow in a catastrophic eruption, obliterating the tourist paradise and home of nearly half a million people.

Both scientists made their claims loudly and directly to a press corps that was eager for sensationalist stories. The authorities, who had hoped that the experts would come and solve the problem of their agitated volcano, couldn't figure out who to believe. Should they take Tazieff's advice and ignore the giant volcano, or should they listen to Brousse and evacuate the island? Tazieff and Brousse battled for control of the crisis with insulting and belligerent jabs, all of which were widely reported in the local media as well as in the French press.

Finally, the Guadeloupe authorities chose not to take any chances and ordered the evacuation of 73,000 people from their homes. It was a grueling move for the residents, who were hauled out in military vehicles and put up in leaky tents with few facilities. Weeks went by, and the volcano did not erupt. Supporting the evacuees had already cost hundreds of millions of dollars and had completely ruined the once-thriving tourist economy. But to the exasperation of the authorities, the scientists still couldn't agree on whether the people should be able to return to their homes.

The situation became such a mess that, finally, an international committee had to be called to resolve the dispute between Tazieff's camp and the group led by Brousse. The committee discovered— unbelievably—that both groups of scientists had come to their conclusions with absolutely *no scientific data*. Tazieff and Brousse, although experienced scientists, were in fact completely without any modern volcano-monitoring equipment and had come to their sweeping proclamations by doing little more than sniffing the volcano.

The analyses by both Tazieff and Brousse were completely worthless. The committee recommended that the volcano immediately be equipped with state-of-the-art monitoring gear to be operated by scientists *trained* to deal with active volcanoes. In response to the turbulent situation with the evacuees, the report stressed the point that individuals and communities always live in areas of some risk, be it weather-, earthquake-, or volcano-related, and advised the authorities to move the people back to their homes.

Soufrière never did have a cataclysmic eruption, but the crisis was a catastrophe economically. The event represented one of the darkest—and most ridiculous—moments in modern volcanology.

Most of the scientists visiting Pasto lived with active volcanoes in their midst and were particularly interested in the story of the horror on Guadeloupe. But it was the Colombians who were especially attentive because Pasto had suffered (albeit on a smaller scale) a similar crisis of scientific rancor and subsequent economic mayhem.

In another attention-grabbing talk later that afternoon, Peter Baxter, a British medical doctor, showed a gruesome slide show and discussed the medical aspects of volcanic eruptions. There were lots of ways to die from a volcano, Baxter told a now wide-awake crowd in Hotel Cuellar's conference room, and even more ways to sustain some pretty nasty injuries. There were mudflows like the one that had buried thousands in Armero. Poison hydrogen sulfide gas could burn your eyes and rot your insides. There were relatively small explosions of ash, steam, and incandescent volcanic blocks that rarely hurt anyone unless there were people actually on top of the volcano—in which case, according to Baxter, about 50 percent survived. But by far the most grisly way to go was death by pyroclastic flow. The blistering heat would cook you until you looked like a charcoal briquette, arms, legs, and facial features intact, at the same time as it solidified your lungs and trachea.

Baxter showed the group a gruesome film he had made of victims of volcanic eruptions: bloody contorted bodies, black flesh melted to bones and faces preserved in charcoal screams. With gallows humor, the crowd nicknamed the grisly expression the "300-degree-Celsius cooked look."

Adams sat uncomfortably through Baxter's film and found himself wondering when the meeting organizers would talk about safety concerns, about the risk of working on Galeras, about medical facilities. He planned on going inside Galeras with Williams's field trip on Thursday. Adams was usually a little worried prior to working on an active volcano. "I was thinking about the work we would do in the volcano, and I was waiting for them to mention some kind of safety plan." Because Adams worked for a government weapons laboratory, he was used to taking safety precautions for all types of work, and active volcanoes were especially serious. But here, where Williams was running the show, it was becoming apparent that safety and emergency issues were low on the list of priorities—if they were on the list at all.

Conspicuously absent from the Galeras workshop were some of the most hardworking and highly regarded scientists in the world—the volcanologists from the U.S. Geological Survey. The group had an unsurpassed wealth of experience, knowledge, and resources and had been the dominant presence at dozens of volcanic crises around the world for years. Several weeks earlier, in reaction to increased terrorism by the Colombian drug cartels, the U.S. State Department had banned survey scientists from attending the Pasto meeting; in fact, their names still appeared on the list of participants. One of those was Bernard Chouet.

Two weeks before the meeting, having just received news of the State Department's ban, Chouet had tossed his meeting invitation in the trash, annoyed that he wouldn't be able to attend. He had not been to Galeras since November 1991, when he had accurately forecast the eruption that damaged the communications towers the following July. He was sorry he would miss the opportunity to educate the international group of scientists about his success at Redoubt, and most recently, at Galeras.

To the three Colombian seismologists at the meeting—Roberto Torres, Diego Gómez, and Fernando Gil—Chouet's absence was a particular blow. In Gil's view, Chouet was an extraordinary scientist. He had been following his work and collaborating with him for years and was about to speak on its significance at the workshop. Gil would present the patterns of long-period seismicity that had shown up before the eruptions of Nevado del Ruiz, and the Pasto observatory's two young seismologists, Gómez and Torres, would report on long-period seismicity at Galeras.

As the two young seismologists began their presentation in Spanish on the seismic history of Galeras, Andy Adams walked out to the lobby for another cigarette. He was getting anxious about working on the volcano, and he wanted to learn as much as he could about Galeras, but he couldn't understand what the Colombian scientists were talking about.

In the chilly, dark meeting room, Gómez and Torres began to tell the fascinating story of how Galeras had acted just before its last eruption. They detailed how, after the lava dome had stopped growing in 1991, the seismicity had calmed down, and Galeras had been completely quiet. Then, on July 11, 1992, a small screw-shaped signal appeared on the laboratory seismographs. Over the next four days, there were just nine long-period events, and Galeras seemed entirely tranquil. Then, on July 16—without further seismic warning—the volcano erupted and blew the dome that had clogged its vent. The explosion registered giant, jagged waveforms on the seismographs. The quakes continued as Galeras readjusted itself, but within a couple of days the volcano's seismic tremors had settled down considerably, and within weeks Galeras again appeared quiet.

But in the minds of Gómez and Torres, this story had a more immediate, and more urgent, moral: they had been seeing the same sorts of data over the past three weeks. Starting on December 23, and recurring each day, Galeras had produced a single long-period event—what they called a *tornillo*. This was virtually the identical pattern that had been seen leading up to the July eruption Chouet had foretold. On the surface, the mountain appeared quiet, sleeping. But Torres, Gómez, and Fernando Gil couldn't help but wonder whether a storm was brewing down below. One thing was certain: Galeras was once again acting suspiciously.

When the official talks were over on Tuesday night, the mood was light and carefree. Andy Adams and his Los Alamos colleagues went to a restaurant for *arroz con pollo* and *cuy*—giant guinea pig cooked on a spit, a Nariño specialty—accompanied by baked potatoes, served whole in their peels and eaten like apples.

On Wednesday morning, the scientists broke into various working groups according to their specialties. Each group spent the day working on a section of the proposal that would later be submitted to the United Nations. Andy Adams and Fraser Goff joined the chemistry group with Stanley Williams.

When their group was finished for the day, Andy Adams and Alfredo Roldán, a Guatemalan chemist working with the Los Alamos team, went to the front desk to sign up for the next day's field trip. They put their names on a trip sheet that read *Geochímica*. There were already over two dozen names on the list for Stanley Williams's field trip, and Fraser Goff decided that he and Gary McMurtry, the fourth Los Alamos team member, would take a different trip. "I didn't want to elbow my way into the fumaroles." His team would stay in Pasto for another three weeks after the meeting. He'd have his time inside Galeras.

Besides joining Williams's trip to the crater, there were four other options listed in the program. Marta Calvache would take a trip to investigate the upper flanks of Galeras where an old Inca trail on the mountainside exposed a beautiful collection of rust red and yellow volcanic deposits. Another group would circumnavigate the volcano and visit the surrounding geology. The fourth group would make gravity measurements with Geoff Brown; and Roberto Torres would take the last group to visit the seismic station on Galeras's summit, and then to the observatory to visit the lab.

By Wednesday night, Calvache collected the field trip sign-up sheets from the hotel lobby. Nearly half the workshop participants had signed up to go to the crater. She didn't like it. Even under the best of circumstances, it was not a good idea to take so many people into an active volcano: now, with a *tornillo* appearing daily on the seismographs, it seemed a particularly dangerous risk—especially to the seismologists. Bernard Chouet had predicted the July 1992 eruption based on *tornillos,* and now the *tornillos* were back.

Fernando Gil, in fact, was very worried about the *tornillos,* and though Muñoz wasn't convinced that they were a dangerous sign, he still had a strange feeling in his gut—that the relative quiet of the volcano was somehow an ominous sign. So Muñoz and Gil met with Calvache, Stanley Williams, and John Stix and voiced their apprehension about the *tornillos* showing up in the seismic data. "We were arguing about what was happening. By January 13 there'd been fifteen *tornillos,* and we had this argument about what it meant," says Gil. Stix and Williams did not seem concerned with the seismic data. The meeting became tense. "Stan Williams stressed that the sulfur dioxide output was low, and therefore he thought the volcano wasn't dangerous," says Muñoz. "The decision was finally made that only the ones that had real reason to enter the crater—the geochemistry and gravimetry groups—would do so."

It was also decided that the trip to the crater would be as brief as possible. "We told them not to stay in the crater too long," Gil says. "We kept telling Stanley not to spend too much time." Even though Williams agreed on downsizing the expedition and making the trip as short as possible, Fernando Gil could not shake his anxiety as the small meeting broke up and the scientists left the hotel. He tried to reassure himself that the *tornillos* didn't foretell an impending eruption, and that the *previous* eruption (in July) had occurred because of the lava dome that plugged Galeras's throat. Now the lava dome was gone, which Gil hoped meant that pressure was being released instead of building in the volcano's throat. Still, he was worried. But he knew Williams would never listen to his concerns: an incident between the two men several years earlier had put them on non-speaking terms.

Back in 1988, after Williams's Nevado del Ruiz workshop, Williams had told Gil that he was going to publish the abstract from Gil's presentation at the workshop in a special volume of the *Journal of Volcanology and Geothermal Research.* Gil told him he couldn't. The work, which was brand-new, was about the theory of long-period seismicity that Bernard Chouet was developing, and neither Chouet nor Gil thought it was ready for publication. They wanted to make sure that the data and analysis were scientifically sound and the results reproducible. Williams had been furious. He screamed at Gil and demanded that Gil let him publish the report. But Gil would not relent—he couldn't even if he had wanted to. The work, after all, belonged to Bernard Chouet. "At the time, we thought Stan and Fernando would come to blows," says Fernando Muñoz.

Five years later, at a meeting supported largely by Williams's grant money from the National Science Foundation, the Arizona State University scientist was clearly running the show. In fact, without the presence of the U.S. Geological Survey volcanologists, including Dave Harlow, John Ewert, and Andy Lockhart, and most important, Bernard Chouet, there wasn't anyone in Pasto with the clout or the experience to stand up to Williams and make the call: *Galeras is dangerous. Do not go into the volcano.*

CHAPTER 9: TO THE MOUNTAIN

**GALERAS, COLOMBIA:
JANUARY 14, 1993, EARLY MORNING**

THE MORNING OF JANUARY 14, 1993, was cool and overcast. The scientists gathered on the front steps of Hotel Cuellar, bundled in warm clothing, looking not unlike groggy, lost sheep in the half-light of morning. It was 7:30, and along the narrow one-way street metal security gates were still pulled shut over closed storefronts. A block-long row of faded four-wheel-drive trucks and a dilapidated brown school bus lined the curb. Each vehicle had been designated for a particular field trip with a handwritten sign—Chemistry, Seismology, Geology—taped to its rear window.

Documenting the day of field trips for the local television station was a two-man team who had been attending the workshop since Monday. It was big news that Galeras had been given the distinction of Volcano of the Decade, and it was something special to see so many foreign experts in Pasto at the same time. The team's reporter had spent three days posing rapid-fire questions in Spanish to the pensive meeting attendees, and he was getting tired of the whole ordeal. Getting the scientists to say anything interesting was nearly impossible and figuring out what they meant was even harder. He wanted someone to make a statement that he could really use. He was looking for some color. That's what viewers like. That's what his editor liked. But when he asked Stanley Williams and John Stix, the two North American experts, "Is the volcano dangerous for Pasto? Is it getting ready to blow?" both said the same thing:

At this time, the volcano is *tranquilo*.

The reporter couldn't help wondering, if Galeras is so tranquil, why are there a hundred scientists in Pasto studying it? He planned to follow Williams's group to the crater. There, he hoped to get some good action shots, and maybe the scientists would be a little more loose-lipped.

Williams and Marta Calvache had planned to have the trucks on the road by 7:30, but things had gotten off-track. The night before, when they had decided to limit Williams's field trip, many of the people who had signed up to go to the crater hadn't been around to hear the change of plans. The scientists wandered along the damp sidewalk, making last-minute decisions on alternative trips. "We got out there early, and there was some confusion," says Andy Adams. "Who was going where, what truck were we supposed to use, that sort of thing."

Pete Hall and Patty Mothes, American husband-and-wife volcanologists living in Ecuador, stood in front of the hotel discussing the change of plans. Mothes had originally wanted to go with Stanley Williams. Instead, she told Hall that she would take the trip with Marta Calvache's group and explore the deposits of ancient pyroclastic flows on the flanks of Galeras. Hall and Mothes gave each other a quick kiss, and Hall left to join the large group boarding the old school bus. Most of those who had been bumped from the crater field trip opted to go on this alternate excursion around the base of the volcano, because the bus was large enough to take up to thirty people. The motor sputtered and the gears scraped loudly as the bus driver inched out and began navigating his way through the town's narrow streets.

One by one, the Jeeps and Toyotas filled with the remaining scientists. Andy Adams and Alfredo Roldán loaded their cumbersome backpacks and assorted equipment into the back of a Landcruiser and climbed into the vehicle's backseat. Nestor García and Igor Menyailov joined them, and the truck took off toward the mountain. Behind them, a white Jeep Cherokee carried Stanley Williams, José Arles Zapata, a chemist from Pasto, and Fabio García, a chemist from Bogotá.

The summit of Galeras sat 5 miles from the center of Pasto as the crow flies, but it took a full hour to make the trip. The road twisted and turned up the volcano's southern flank and was strewn with rocks and boulders, offering the scientists—who soon gave up trying to take notes in the moving vehicles—an extremely bumpy ride. By the time they reached 12,000 feet, the mist had been replaced by a dense white cloud, and there was nothing to see. Andy Adams and Alfredo Roldán grumbled about the view, and hearing the complaint, the truck's driver happily relayed *Pastuso* legend.

"Galeras is shy," he told his passengers. "He doesn't care too much for visitors. If the day is clear and you start up the mountain, Galeras will know you're coming and will hide himself in the clouds."

By 9 o'clock, the groups led by Williams, Calvache, and Roberto Torres arrived at the summit, where they would meet before setting off on their individual ventures. There were more than thirty scientists on the rim of the caldera. García, Williams, and Zapata piled out of their white Jeep and walked toward a building emerging eerily from the fog. It was a large, two-room structure with four 30-foot radio towers protruding from its roof: the police station. A military policeman appeared in the doorway and lifted his head to the group in a nonchalant greeting. García followed Williams and Zapata down a concrete pathway littered with irregular wooden planks and bags of cement that ran along the front of the police station. A plain stone cross rose from a pile of rock, giving a ghostly, cemetery-like feeling to the top of Galeras. García looked around uncomfortably. "It was impossible to see anything clearly that was more than ten meters away," he recalls. Beyond that, everything disappeared.

On the police station's cement path, the group led by Marta Calvache gathered around her. It was disappointing for many of the foreign scientists, knowing that they were standing on top of an active volcano but unable to see anything. But many who had originally planned to go with Stanley Williams were just as happy now that they weren't. "There was absolutely nothing to see," says Patty Mothes.

And because of the dismal visibility, there was no reason to hike farther east along the ridge, so the scientists mingled in small groups around the police station. Calvache tried to give them a sense of what they'd be seeing if Galeras had been more cooperative. They were standing on the southeast rim of a mile-wide amphitheater called a caldera, she told them. The caldera forms a 500-foot-deep bowl, a quarter of which is completely open to the west, giving the volcano the appearance of a horseshoe from an aerial view. The ridge where the police station stands is 40 feet across—the widest portion of the caldera rim. Traveling around the caldera, the ridge becomes very narrow and impossible to walk across.

Setsuya Nakada listened politely. He had come a long way to look at Galeras, and now there was nothing to see. Nakada was a petrologist—a scientist who studies rocks under a microscope and can trace the history of a volcano by the way that minerals and textures of the rocks change from one eruption to the next. "At first I wanted to go down and take a sample from the lava dome that had blown out in the July eruption, but I checked the photograph and I thought it looked very dangerous. The rocks were very oxidized, and to take a fresh sample, you would have to enter the crater."

To Nakada, the crater looked like an incredibly bad place to be, and he decided that there was no way he would go near it. Fortunately, the previous eruption had blown the lava dome to millions of pieces and deposited plenty of fresh rocks on the caldera rim where he stood. He picked up a fist-sized chunk of gray andesite and turned toward his colleague, Tad Ui. "This is good enough for me," he chuckled, examining the rock's white crystal flecks through a dime-sized magnifying lens.

The rest of the group surrounding Marta Calvache was growing bored. After asking a few questions, they stood and looked into the white mist, waiting to see if the clouds would clear. Along the caldera rim to the east, Williams and his group were gathered, and closer to the police station, the reporter was interviewing Patty Mothes. He had been interviewing her for more than fifteen minutes. The video shows Mothes standing on the exposed ridge, her floppy hat dancing around her face, her cheeks flushed in the chilling wind.

"When will there be an eruption?" asked the reporter.

"It's impossible to say."

"Will there be an eruption in five years?"

"We can't say absolutely when Galeras will erupt. There is a probability that it could."

"Is it dangerous for Pasto?"

"Right now, the volcano is quiet, but there is a likelihood for a future eruption."

Mothes was tired of the endless questioning but did her best to mask her frustration. She explained that the scientists at the workshop had come to use Galeras as a real-life laboratory. Learning about Galeras's personality would help them understand what the volcano had in store for Pasto, and it would also offer them information about similar volcanoes around the world. Then she gave a lengthy explanation of why people needed to get a handle on understanding volcanic hazards.

The journalist fidgeted with his microphone, growing impatient.

"So when is it going to blow?"

Mothes gritted her teeth. "It's *impossible* to tell."

"Is Galeras an active volcano?"

"Yes. Definitely active."

The wind lashed across the ridge as Mothes returned to the

group gathered around the police station. Thick fog continued to fill the caldera.

Alfredo Roldán stood quietly on the summit ridge next to Andy Adams, smoking a cigarette. They both wore white hard hats and steel-toed boots, and Adams was dressed in brown fire-resistant coveralls. The 5-foot-11 Adams had put on weight in the past few years, and the brown suit, along with his mustache, made him look a bit like a walrus. Stanley Williams gestured to the two and laughed.

"He made fun of us, and they all laughed because we were all dressed up with our safety gear," says Adams. He was miffed. No one else had a hard hat, and most of them didn't even have gas masks. "Some of them had never been to an active volcano," Adams says.

Adams's expression had started to sour. He was used to working with people who took safety in an active volcano a lot more seriously; no one had ever made fun of him for wearing a hard hat. "I thought, 'Some expert,' " Adams says. Williams acted like they were going on some kind of picnic. But in the hierarchy of the scientific research community, Adams (and his master's degree) was outranked by the Ph.D. from Arizona State University. "I was a peon. Stan was the senior scientist." So Adams glowered and kept his mouth shut. Roldán, as usual, was quiet. His face never gave away what he was thinking. "I couldn't believe that Stan was laughing because we were wearing hard hats. I always wear one, and I couldn't believe that only Andy and I had them," he says.

Farther east along the ridge, Geoff Brown was fiddling with his gravity detector. Brown was a soft-spoken British geophysicist with long white hair and matching beard. He had brought with him an aluminum box the size of a car battery that he hoped could be used to track molten magma as it rose to the surface of a volcanic vent. The instrument detected infinitesimal changes in gravity, and because magma is lighter than solid rock, Brown was hoping to tell by testing the gravity field beneath Galeras if there was an eruption on the way. It was a long shot; the method was very new and had no proven track record. But Brown was optimistic that he would eventually have success.

Joining Brown in his work and watching him inspect his gravity box were two Colombians: Carlos Trujillo, a local *Pastuso* who taught civil engineering at the University of Nariño in Pasto, and Fernando Cuenca, a young geophysicist from Bogotá who had just finished graduate work in Russia. For Trujillo, the meeting was a thrill, and so was the chance to work with a distinguished English scientist. The university where Trujillo taught didn't have a geology program, but he had always loved the volcano and this was his chance to meet some of the biggest names in volcanology.

Geoff Brown sat his gravimeter on a makeshift stand, turned and twisted the dials, put his nose right up to the instrument, checked the readings, and took a few notes. Brown spoke Spanish marginally well and happily agreed when the reporter requested an interview. The video shows the exchange:

"Professor Brown, why do you study gravity at the volcano?"

"It's important to know the volcano well and to know the changes in gravity, among other things, in order to predict eruptions in the future." His voice was warm, and his eyes crinkled as he smiled. His wispy hair blended into the white backdrop of fog behind him.

"When will Galeras erupt?"

"I don't know exactly. It's necessary to know the volcano much better for there to be a prediction. It's not possible to predict an eruption right now. Possibly in ten years, possibly a hundred. I don't know."

Two thousand feet below the summit, Marta Calvache's group stopped for lunch. At 1 o'clock, Calvache gave the scientists the option of going back to Pasto or taking a half-mile hike to look at an outcrop of pyroclastic deposits. Several wanted to return to Pasto and they piled into a truck and were taken down the mountain by an observatory driver.

Calvache led the six remaining people in her party north along a narrow foot trail that wrapped around Galeras. It was an old Inca trail that the scientists had named Camino Real, and it led to some of the best rock outcrops on the volcano. They walked for a quarter-mile along the path, thick with tall grasses on either side, until they arrived at a 10-foot wall of layered red and yellow crumbling rock.

The scientists took out blue-handled rock hammers and beat on the outcrop until pieces fell into their hands. With small magnifying lenses, they inspected the crumbs, the colors, and the small black-and-white flecks in the red matrix. To the geologists, it all had meaning, every mineral, every texture, every rock layer—it was all part of the history of Galeras. Calvache explained how the million-year-old stratovolcano had built itself into a grand mountain with layers of lava flows and pyroclastic flows, how its fragile sides collapsed, and how it began to grow again. The volcano had climbed to the sky and fallen to the earth several times in its long life, and now the small cone on the caldera floor was Galeras's resurrection.

Patty Mothes had been to Galeras before, and while the others inspected the rocks, she turned away and looked to the east. The day was now clear, and she could see Pasto down below, its tall white churches and neat rows of streets bright in the sunlight. It was a lovely view—far better, she knew, than what they could see from the volcano's summit, which was still covered in clouds.

CHAPTER 10: THE DESCENT INTO GALERAS

GALERAS, COLOMBIA:
JANUARY 14, 1993, 9:30 A.M.

IT WAS JUST AFTER 9:30 in the morning when Stanley Williams's group began its descent into the dense fog that filled the caldera of Galeras. Freezing wind whipped over the ridge, chilling the gathered scientists, and they were anxious to get down below the summit where they would have some protection from the cold. On the southeast side of the volcano was the entrance into the caldera: a yellow nylon rope fixed to a rusty cleat cemented into the rocky ridge. The entry point, 100 yards east of the police station, was chosen because the slope here was a mere 50 degrees, while most of the amphitheater's inner wall was a near-vertical face of crumbling rock.

Williams explained that they would go down the rope one at a time, to prevent rocks from dislodging onto the person below. The rope was 100 feet long and would take them down the steepest portion of the descent. From there, the slope was made completely of scree, marble-sized rubble shed from the caldera wall, and much easier to navigate—at least on the way down.

As they waited their turn, the scientists carried on overlapping conversations in Spanish and English while they filled heavy packs to capacity with equipment. There were glass vials and flasks full of chemicals, titanium tubing and metal wiring, small portable computers, drinking water, and ice coolers for keeping samples cold. Some of the equipment was the same, but this trip into Galeras was much different than the expeditions by the U.S. Geological Survey in 1989. There were thirteen scientists in Stanley Williams's group, and among them, only two wore hard hats, only two had full-face gas masks. Only one person in the entire party carried a radio. No one was positioned on the summit to receive a radio signal from the Pasto observatory, and back at the lab, a student, not a seismologist, was positioned to watch the seismometers. No one carried emergency medical supplies. And this time, a pair of reporters, untrained and totally unprepared to go into an active volcano, were invited to follow the group to the crater.

José Arles Zapata, who was representing the Pasto observatory on the field trip, readied himself to go down the caldera wall. Next to Marta Calvache, Zapata knew Galeras's crater better than anyone. He had been there nearly every week for the past two years collecting samples from the fumaroles. He knew it was dangerous work—he had just missed, by a few days, getting blown up in the eruption in July 1992—and Zapata had a reputation for conscientiousness. As was his routine, he made a quick radio check with the observatory.

"José Arles called me by radio from the top of the rim," recalls Adriana Ortega, a geophysicist who was watching the seismometers at the observatory. "It was about nine-thirty in the morning, and he asked for a report on the seismicity. I told him everything was quiet."

Zapata stuck the radio back into his pocket, tightened the straps on his pack, and, using the rope, began the descent along the caldera wall. He navigated the steep escarpment with ease, making it to the end of the rope in less than a minute.

Igor Menyailov went next. Williams had told him that there were gases hotter than 300 degrees centigrade coming out of the crater of Galeras. It sounded incredible, and Menyailov was itching to get into the crater to see if it was true. Williams followed the Russian, and as he lowered himself down the slope, he told the rest of the group that he would wait for them at the bottom. In less than a minute, he disappeared into the fog. The wind continued its strong assault on the ridge, and the scientists pushed their hands deep into jacket pockets. Williams called up that he had reached the bottom of the slope several minutes later.

Geoff Brown was next. The British scientist was anxious to get more familiar with his new gravity instrument and be sure it was working properly—he had been out working the previous day collecting gravity data from the center of Pasto to the top of Galeras, and the numbers his device had recorded had seemed suspect. Carlos Trujillo and Fernando Cuenca helped Brown pack the gravimeter into a special backpack, and when the route was clear, Brown descended below the caldera rim. Trujillo went next, disappearing down into the mist, followed by Cuenca.

As the group waiting on the caldera rim grew smaller, Mike Conway and Andrew Macfarlane, two American scientists, hiked toward the cleat embedded into the cement. "I was glad to have the rope because I had the heaviest pack of the three of us [about 50 pounds]," recalled Macfarlane in a document he wrote several weeks after the eruption. They were planning to set up equipment that would monitor the temperature of the fumaroles—most of it they'd be leaving in the crater for future monitoring. Macfarlane was relieved that he wouldn't have to carry the heavy load back up the slope.

Conway made it down and then shouted up to Macfarlane that it was safe to follow. It took him about ten minutes more, and then the two joined Williams and the others at the bottom. Alfredo Roldán arrived next, followed by Andy Adams. Adams, overweight and struggling against the effects of a two-pack-a-day habit, moved slowly, further encumbered by his thick safety suit.

The visibility was better down inside the caldera than it had been above, and the others watched Igor Menyailov's bright red jacket and short white hair as he broke from the team and raced across the flat caldera moat toward Galeras's steaming cone. Rocks the size of compact cars dwarfed the Russian as he ran up the rubble slope. It was the first time Andy Adams had actually seen the huge brown pile of rock, 2,000 feet wide at its base and 450 feet high, that made Galeras's cone. In its center, a vast crater 400 feet across marked the mouth of the volcano. The rim of the crater was uneven, jagged, and scalloped from centuries of explosions. Adams worried that he'd be lagging behind all day. "I looked at the cone—I was already tired from making it down the caldera wall—and I thought, 'Some piece of cake.' "

Cottony clouds of steam rose from the crater, disappearing into the gusty wind 200 feet into the atmosphere. From the fumaroles on the western slope of the cone, steam shot out as if from the spout of a boiling teakettle. Williams led the group toward the cone, and Adams stopped to catch his breath. After several minutes, he continued walking, but the trip was much harder than he had expected. In the slide shows given at the workshop, the cone looked small and inviting; now it looked gargantuan. He walked slowly, breathing hard in the thin air, staring at his feet and trying not to trip on the loose rock. "For a while, Alfredo waited for me, then he went on. Then the whole group was at the top of the crater waiting for me. I was slow," says Adams. Alfredo Roldán had walked ahead and caught up to José Arles Zapata, who was decked out in white pants, a yellow jacket, and a stylish scarf—a flamboyant field costume for a geologist.

Having worked on the wildly explosive Guatemalan volcano, Pacaya, Roldán's perspective on Galeras was quite different than Adams's. Roldán remembers: "I joked with Zapata, I said, 'Hey, José Arles, this isn't an *active* volcano. If you want an *active* volcano, then you should see Pacaya.' " Zapata laughed and told Roldán to wait until the clouds cleared—*then* he'd see that Galeras was definitely active. "Of course, I knew that Galeras was active," Roldán says. But it still seemed like a pretty puny volcano compared to Pacaya.

The cold felt good, and the mood was light as the group trudged over the rough ground. It was clearing nicely inside the caldera, but menacing clouds still blanketed the caldera rim. The weather at Galeras could change in minutes—and it did, over and over, thanks particularly to the open western wall of Galeras's caldera, which gave clouds a free path. Then a stiff breeze riding the same course would push its way into the amphitheater, displace the frothy clouds, and again it would be clear.

Back in town at the Pasto observatory, Adriana Ortega watched the seismic data. A *tornillo* showed up on the seismograms. It was 9:45 A.M. The small squiggles didn't look too threatening, but the policy was to let the people working in the crater know when something showed up on the seismographs. "I called José Arles on the radio to tell him, but he didn't answer," she says. "I called a couple of times, then Roberto called in."

On the caldera rim, Roberto Torres and the seismology field trip group were 500 yards west of the police station and almost to the radio towers where the seismic station was installed. The group moved slowly up the steep slope, unaccustomed to the altitude. Torres had heard Ortega calling on the radio, and he wondered why Zapata wasn't answering. Torres took his radio out of his pocket.

"José Arles, come in," Torres called.

Within seconds, Zapata called back. "José Arles here."

"The observatory is calling you. Are you picking up the transmission?"

"No, I didn't get the call."

At the observatory, Adriana Ortega heard Roberto Torres calling to Zapata but couldn't hear any answer from Zapata directly. "I thought there must be a problem with José Arles's radio," she says. "So I called to Roberto and told him to tell José Arles that there is a small *tornillo* in the seismograph. It was small, but I wanted him to know."

Torres relayed the information. "José, there is a small *tornillo* in the seismograph. Adriana wants you to know."

"Okay, Roberto. I read you. Tell the observatory that we'll make the trip to the crater quick and get out of here as soon as possible."

Torres relayed the message back to Ortega, put his radio back in his pocket, and looked down into the volcano. The appearance of yet another *tornillo,* coming on the heels of the conversation with the meeting organizers the night before was an ominous sign. Torres checked his watch, satisfied that Williams and Zapata would hurry things along.

Williams and the other chemists stood at the top of the crater rim and looked down into the steaming crater. As always, the air stank with sulfuric acid, and the view was still somewhat obscured by the ever-shifting clouds. Occasionally the fog would clear briefly, revealing the crater of Galeras as a sheer-walled depression 100 feet deep. Its floor was a massive pile of brown and yellow stones rotted from the acid, and several thick fountains of billowing steam gushed up in a low hissing roar. For ten minutes, they stood silently, watching the volcano breathe.

Luis LeMarie, a hardy Ecuadorian chemist from Quito, went back down the cone to walk with Andy Adams, and the two gradually made it up to the crater rim. By the time they joined the group, the sun was shining, the sky had cleared, and they could finally see in a full circle around them. The caldera walls were incredibly steep, built from ancient brown and gray lava flows that had been streaked red and yellow by acidic volcanic fluids percolating through the rocks.

Geoff Brown and the gravity team were already out of sight. They had left the main group and were circumnavigating the crater, moving clockwise from the south and taking measurements with the gravity detector along the way.

Stanley Williams stood on the crater rim, casually overseeing the other scientists and directing them around the cone. Igor Menyailov had gone off on his own, and the group could see his slim figure to the west, crouched down next to a 30-foot column of white steam spewing from the cone's western slope. José Arles Zapata pointed toward the Russian and explained to the foreigners that all of the fumaroles have names. The three where Menyailov was working were called Deformes—Deformed. The west flank of Galeras's cone bulged under the pressure of three gushing fountains of steam, hence the name.

Andy Adams knew that the chemistry of the gas coming out of Deformes would be identical to what was coming out of the volcano's throat. "I was relieved," he says. "There wasn't any need to go into the crater." As it was, he and Alfredo Roldán could stay on the cone and take samples of the gases, a much safer thing to do, he thought.

The scientists began to divide into smaller groups. Fabio García, the chemist from Bogotá, joined Andy Adams and Alfredo Roldán taking temperature measurements. LeMarie joined the Americans Andrew Macfarlane and Mike Conway, who were collecting gas samples, while Williams stayed on the rim and periodically checked in with each group.

Igor Menyailov was excited to get into the crater, but he took his samples at Deformes first, sticking a glass tube into the steaming ground. Menyailov worked using a deadly contraption called a Giggenbach flask that contained liquid cadmium. Most scientists had given up the method because it was too dangerous to carry the poisonous substance, which had to be contained in glass tubing. The Russian also performed his sampling in the hottest and most dangerous parts of the volcano, wore no gas mask, and was often seen enveloped in toxic fumes smoking a cigarette. Menyailov's work style shocked many scientists, but Stanley Williams admired the Russian's tough and crazy image and often commented that Menyailov was *his* kind of scientist.

In 1993, the connection between volcanic gases and volcanic eruptions was a hotly contested mystery. Because rising magma releases sulfur dioxide, some scientists, including Stanley Williams, believed that if the fumaroles spewed large amounts of sulfur, it meant that magma was rising to the surface, burping up sulfur-filled gas, and that the volcano would soon eject the magma as thick lava, possibly producing pyroclastic flows. In the case of Nevado del Ruiz, after the 1985 eruption, the fumaroles roaring from the crater contained abundant sulfur dioxide, and from that data, Williams predicted that a colossal eruption, much deadlier than the one that wiped out Armero, was on the way.

But the eruption never came. By 1993, at the start of the Galeras workshop, there had never been a successful prediction based on gas data. Nearly all the world's top volcanologists were convinced that gas data—in and of itself—was not a useful tool for prediction, but it was a mission that Williams and a few others in the volcanology community, including Igor Menyailov, doggedly pursued.

The wind was gusty, and the gases, heavy with acid, were blowing in all directions. The others walked around the cone to Deformes to watch Menyailov work. Andrew Macfarlane could taste the gas in his mouth, so he put on his gas mask. Mike Conway and Luis LeMarie didn't have masks, and their eyes and mouths burned in the acid steam. Macfarlane decided to wait to start his sampling until Menyailov was finished with Deformes. "Being new to this kind of study, I also wanted to observe the technique of more experienced workers," Macfarlane recalls. After a half hour, Menyailov was finished. He gathered his flask and other materials and headed for Galeras's inner crater.

Nestor García hadn't planned to do any sampling on the field trip, but he couldn't resist joining the crazy Russian in the mouth of Galeras. He borrowed a thermocouple (temperature-measuring device) from Andrew Macfarlane and hurried after Menyailov.

It was 12 noon.

José Arles Zapata wanted to check with the observatory while the seismologists, including Fernando Gil and Bruno Martinelli, were still there, just in case there was anything to report. He knew they'd all be going to lunch soon and would probably be gone for a while.

"Zapata to Pasto. Come in Pasto." He held the radio and waited for a response. None came.

He tried again. "Come in, Pasto."

Still no answer.

Roberto Torres, the seismologist from the observatory, was still hiking down the slope of the caldera rim with the six people in his field trip group. He heard Zapata calling on the radio, and he wondered why the observatory wasn't answering.

Torres took out his radio and called the observatory. "Adriana, are you there?"

"Yes, Roberto. What is it?"

"José Arles is trying to reach you. Are you picking up his transmission?"

"No. I haven't heard from him."

In the caldera, on the slope of the cone, José Arles Zapata had put the radio back in his pocket, wondering why he wasn't getting through to the observatory. Then he heard Torres crackling through.

"Roberto, José Arles here."

From the caldera rim, Torres could barely make out the small shapes of people working near the crater. He saw a bright yellow jacket next to Deformes and knew it was Zapata.

"José Arles asked me if I could get him a report on the activity of Galeras," Torres says. "And the observatory told me that everything was calm, Galeras had been quiet since the one *tornillo* in the morning." Torres gave José Arles the report, and Zapata told him that the group was finishing their trip and would be out of the volcano very soon.

But on the cone of Galeras, there was no sense of urgency to finish working. According to Andy Adams, Alfredo Roldán, and Andrew Macfarlane, Stanley Williams never tried to hurry out the group or press them to be quick about their sampling. Instead, he made casual conversation with the scientists and encouraged a leisurely pace.

All the while, an immense pressure was building within Galeras.

* * *

At Deformes, the steaming fumarole 100 feet below the crater rim on the western slope of the cone, Andrew Macfarlane and Mike Conway unpacked their equipment and got ready to take gas samples where Menyailov had just been working. Luis LeMarie had joined them and was taking the temperature of the fumarole with a thermocouple. Macfarlane took out his field notebook and wrote down the number that the device was showing: *Deformes fumarolas, 201° C.*

As Macfarlane set up the testing equipment, Williams stood nearby, talking with a man and two teenage boys whom Macfarlane hadn't noticed before. The three didn't seem to be working, so Macfarlane figured that they were tourists who had climbed down the yellow rope after them.

The fumarole roared and spewed steam robustly, the cloud reaching 30 feet high before dissipating into the wind. "Mike and I were engulfed in sulfurous vapor for much of the time we were sampling," Macfarlane says. Gusty winds blew the gas in all directions, and Conway and Macfarlane tried to dodge the thick acidic plumes.

Macfarlane looked at his new hiking boots and noticed that the metal eyelets were already corroding in the hot acidic steam that emanated through the rotten clay stones he was standing on. After a half hour, Conway and Macfarlane, along with LeMarie, began to wrap up their work at Deformes. Macfarlane checked his watch.

It was 12:30.

Adams, Roldán, and Fabio García had come around the cone from the west and were talking with Geoff Brown's team. Fernando Cuenca, García's colleague from Bogotá, took García aside. "He got my attention, and he told me that the gravity apparatus wasn't working," says García. "They were getting values that weren't any good, and Fernando thought something was wrong." Nonetheless, they still planned on working the rest of the day to finish—even if the data weren't any good.

Adams, Roldán, and García continued counterclockwise around the crater rim until they reached the group that were still taking samples at Deformes fumarolas. When they were gathered close enough for everyone to hear, Williams said that it was getting late and recommended that some of the group begin to leave the caldera and get a head start.

Andy Adams had taken enough ribbing to get the hint. "I was sure that Stan meant me," he says. But he was exhausted and had no problem getting an early start on the hike. Sweat dripped down his temples to his neck and from his forehead into his eyes. He wished he could take off his hard hat, but there was no room in his pack to put it.

Alfredo Roldán prepared to leave with Adams, and the two were joined by Fabio García and Carlos Estrada, the driver from the observatory.

It was 12:45 P.M.

Andrew Macfarlane and Mike Conway had more work to do after they finished taking gas samples. Their main mission wasn't to collect gas; they planned to run a wire attached to a thermocouple linking all of the fumaroles in the crater to continuously keep track of temperature changes.

"After we finished taking our sample, we got out of the gas and caught our breath for a little bit," recalls Macfarlane. "Stan was teasing us about how we were real volcanologists now because we had 'sucked the gas' at the fumaroles—like a rite of passage. He also said that Igor and Nestor had gone down to the crater floor, which was tough going, but easier than he remembered from the last time he had been there."

The clouds above the caldera rim were starting to clear slightly, and the open amphitheater offered a spectacular view of the Azufrado Valley and the rolling green hillsides of the surrounding countryside to the west. The sun warmed the caldera, and the temperature rose to the fifties. As Adams and the others headed back down the cone, en route to exiting the volcano, Macfarlane, Conway, and LeMarie gathered their gear and headed in the opposite direction, uphill toward the crater rim. They examined the positions of the fumaroles inside the crater and found that they were much farther apart than they had thought. Conway didn't think they would have enough wire to link as many of the fumaroles as they had planned.

The two Americans discussed the problem and came to the conclusion that rather than go into the crater today, they would cache the data logger, thermocouples, and gas sampling vials and go back to Pasto and discuss what to do. They left the equipment in a shallow hole and put some large rocks on top to protect it in case it became windy overnight.

"Stan remarked that we could either start on some new exploration and then maybe be late getting back to the caldera rim or leave now and have an easier time going back," recalls Macfarlane. The three chose to leave, so they joined Williams and José Arles Zapata and started walking counterclockwise back toward the south side of the cone.

They soon met Geoff Brown, Fernando Cuenca, and Carlos Trujillo, who were making their way clockwise around the crater. Brown and Williams stopped and spoke briefly, and Andrew Macfarlane took out his camera and began to take pictures back toward the crater. The clouds had cleared, and he could see Igor Menyailov and Nestor García near the northwest wall of the volcano's throat.

Williams and Brown finished talking, and the gravity team continued north into the fog that now rose above the crater's edge. The rim along the crater was very narrow, and Macfarlane watched as Cuenca and Trujillo tried unsuccessfully to keep up with the much older British scientist. Then they disappeared into the mist.

The wind blew strongly and began to push the clouds out of the crater again, offering a brief clear view of the crater. The sun caught the various minerals that were flowing in the columns of steam, and they appeared surrounded by halos of pink, yellow, and green. "Igor was no longer near the wall of the crater floor," recalls Macfarlane, "but was out near the middle having a cigarette and enjoying the view. He looked like he had finished his sampling and was feeling satisfied with everything."

As Williams and the others left the crater rim and started to walk down the slope, the rumbling sound of falling rocks seemed to come from the direction of the crater. "There were three moderate-sized rock slides within the space of a minute or so. They were clearly audible, but I couldn't see where they had taken place. I asked Stan about them, and he said he thought there might be some microseismic activity, but he didn't seem really alarmed," recalls Macfarlane.

To Macfarlane, it sounded like the rockfalls that he had heard when hiking on dormant volcanoes in the Cascades of Washington and Oregon. He told himself not to worry.

The group continued down the southeast side of the cone. Ahead of them, the three local hikers navigated the rocky slope. They were dressed in wool sweaters and cheap tennis shoes. Stanley Williams pointed at the three, to whom he had been talking earlier. They were locals from Pasto—a father with his teenage son and the boy's friend.

"Sometimes I feel silly hiking up volcanoes with a thousand dollars' worth of mountaineering gear, while the locals just climb up in whatever jeans and sneakers they have," he said to Macfarlane, who was several steps behind. They both laughed. It was true. The Americans' flashy Gore-Tex and synthetic fleece stood out like an Armani suit at a tractor pull.

It was 1:42 P.M.

The trip back had taken twice as long as it should have. Alfredo Roldán, Fabio García, and Carlos Estrada had been waiting on Andy Adams, and it wasn't until 1:30 that they finally approached the base of the steep caldera wall.

Roldán quit walking and stood still. He heard a low rumbling sound. He'd heard the noise before. It sounded like what he called *retumbos*—low and deep, similar to a rock slide, but a lower-pitched reverberation. "I asked Fabio García what the sound was," says Roldán. "He told me it's just falling rocks. Carlos Estrada said that it was normal for Galeras."

Roldán wasn't convinced. At Pacaya, the Guatemalan volcano,

he had heard the same sound before—just minutes before the volcano erupted.

They started up the slope of loose rubble. Roldán decided that he had waited long enough for Adams and took off up the caldera wall. He moved fairly quickly and made it to the bottom of the climbing rope. The jagged slope was difficult, and even with the rope, and he had to stop several times to catch his breath. When he finally reached the top he saw the journalist and the cameraman, who had evidently left the crater earlier and were now waiting for him at the summit.

Shit, he thought to himself. *Now I'm trapped.*

He hadn't had any water since they had started out that morning; he was dying of thirst and breathing heavily. But the journalists didn't give him a chance to stop and rest. The reporter stuck the microphone in Roldán's face. The video shows Roldán catching his breath as he tries to answer the reporter's questions:

"Engineer Roldán from Guatemala, tell us, what are your impressions about the volcano?"

Roldán looked at the ground for a few seconds, breathing heavily, and then back up at the journalist. "It's very interesting to study it," he said, hoping his answer would be sufficient.

The reporter tried again. "Now that you have been in the crater, what do you think about the imminent risk of Galeras? What are your recommendations?"

Roldán thought about what he could say to get rid of them. "I think that it's a convenient volcano to conduct geothermal research in."

"Listen," the journalist said impatiently, while the cameraman momentarily stopped recording, "I'm going to ask you a question, and you will give me a simple answer. And we will record it. Is that okay?"

"All right," Roldán grumbled.

"Engineer Roldán, when do you think that Galeras will erupt?"

It figures, thought Roldán. *The typical question.*

The camera came back on to tape Roldán's answer. He tried to keep it simple.

"It could be a year from now, it could be a month from now." He stopped and took a deep breath. "Or it could be next week."

Suddenly, the videotape goes blank. On the audiotape that continued to run is the sound of an enormous explosion, and the bloodcurdling screams of the terrified newsmen.

CHAPTER 11: THE ERUPTION

ON THE CONE OF GALERAS, the scientists froze. Beneath the crater, water had turned to steam, creating an immense amount of pressure impossible to confine, and the volcano exploded. The eruption sounded like a horrifying clap of thunder. For a second, no one was sure what had happened. "Our first reaction was to turn back and look," recalls Luis LeMarie. A thick surge of black ash shot from Galeras's throat and incandescent boulders exploded into the sky in a hellish firestorm.

Someone screamed, "It's the volcano!"

Stanley Williams and José Arles Zapata took off down the rocky slope with Macfarlane right behind. Within seconds, force from the explosion heaved meter-wide boulders hundreds of feet into the air, and smaller rocks shot thousands of feet into the clouds. In seconds, boulders rained from the mist and crashed to the ground— fragments of glowing shrapnel exploded in all directions in ear-shattering bursts.

A golf ball–sized rock clipped Macfarlane above his left eye, sending him into a contorted cartwheel that dropped him into a pile of fiery shards. He forced himself up and continued to run, the ground hissing beneath his feet. Sharp fragments pummeled the backs of his legs like flying daggers on a horizontal trajectory, as boulders shattered on impact.

Seconds after, Williams was knocked to the ground when a rock struck his head. Just below, Zapata ran as fast as he could, hurdling over glowing boulders until a sharp rock sliced through the sky and lodged into the back of his skull, slamming him face-first into the burning rubble.

The local man and two boys ran for their lives. Rocks collided with the earth all around them, detonating on impact. Within seconds all three were felled—first the man, then the two boys, their skulls hammered unmercifully. They collapsed to the ground, their clothing on fire, volcanic blocks pummeling their bodies.

At the base of the caldera wall, 2,000 feet from the volcano's throat, Andy Adams dove behind a boulder "and I tried to crawl under my hard hat," he recalled. From the crater, rocks that were originally shot vertically were losing velocity and then dropping from the sky. The force of the continuing eruption blew the falling stones on a horizontal trajectory, creating a glowing fountain 4,000 feet in diameter, 360 degrees around the cone. A burst of stones up to a foot across hit the wall above Adams, shattered, then cascaded down on top of him in a fiery avalanche. He felt the crack of rocks pelting his hard hat and burning his neck and held his pack over his head for more protection. Trembling, he took out his gas mask and put it on.

Above, on the caldera wall, Carlos Estrada and Fabio García grabbed desperately to the climbing rope with one hand, and with the other, tracked the arc of approaching projectiles like outfielders, dodging them as they struck the slope.

The journalist on the summit screamed when they heard the explosion. The reporter fell to the ground and pulled the cameraman down with him, then yelled to Alfredo Roldán as rocks slammed into the summit.

"What's happening? Jesus Christ! What the hell is happening?"

Roldán ran back to the south slope of the caldera rim where he could no longer see the volcano's cone and got on his knees. "I yelled at the journalists to get back on the ridge and get down," Roldán says.

The newsmen ran frantically toward Roldán and tackled him on the slope. Roldán wanted to take a picture—the fantastic black eruption cloud was now spiraling with billows of gray-white—but the reporter and cameraman were on either side of him still screaming and pinning him down so he couldn't free his camera. Rocks pelted the summit like giant, crackling hailstones.

"I told them to calm down," Roldán says. "I told them, 'You're not going to die!' "

Just then, Roldán looked up. "I saw a glowing rock 20 centimeters across coming for the journalist's head," he says. He shoved the journalist hard, and the rock flew past, missing his skull by inches. Moments later, another slammed into Roldán's hard hat, knocking him sideways.

The mountain exploded three more times in rapid-fire succession. A mass of dead black ash—andesite rock pulverized by the eruption's megaton force—fired into the sky in a hideous peal.

Down on the cone, Stanley Williams struggled unsteadily to his feet and tried his best to run. Within seconds he was felled again, as exploding rocks struck his legs, splitting his shin and setting his clothes on fire.

Andrew Macfarlane, farther behind, ran past José Arles Zapata. The young Colombian's body was covered in blood, still-burning rocks embedded in his flesh. Macfarlane stopped for an instant but knew there was nothing he could do: Zapata was dead.

Higher on the cone, Luis LeMarie and Mike Conway took cover near the crater rim. "People were dropping like flies," Conway says. "From that point of time we're crawling around—Luis LeMarie and

I—and the top of the rim was protecting us so we were huddled up there for a minute."

"The rocks started falling down even before we were able to reach the earth. I felt a flow of ash, very hot, passing over my head and hitting my body," recalls LeMarie.

Conway saw Andrew Macfarlane below him on the slope of the cone and yelled at him to get behind a boulder and hide.

Macfarlane heard Conway, found a 2-foot-high boulder, and ducked behind it. The largest rocks had quit falling, but missiles the size of footballs still rained, exploding on impact. The smaller rocks, having been ejected so high into the atmosphere by the force of the eruption, seemed to be coming straight down instead of on a trajectory from the crater, and Macfarlane didn't think the boulder could protect him. He looked up and saw Conway tumble down the slope toward him, arms flailing wildly.

There were so many rocks falling that Macfarlane knew there was no chance to see them all. "By then I fully expected to die at any moment," Macfarlane recalls. "I decided to just keep going rather than hide and wait, and hope that somehow I would be missed by the bombardment." Blood from the wound on his head filled his eye and made it difficult to see. He got only a few yards before he lost his footing and collapsed. The fall sent him toppling over a large boulder, and he landed headfirst, facing back toward the crater. Five feet higher on the slope lay Stanley Williams.

"My leg is broken!" Williams yelled to Macfarlane over the roar of the purging volcano. "My leg is broken! It's severed!" Both bones in his lower left leg were broken cleanly. His booted foot hung limply, the lower half of his shin at a 90-degree angle from the upper half. Blood ran from underneath Williams's knit cap.

Macfarlane wanted to help. "The only thing I could think of that could help him would be to pick him up and carry him, and I reached out to him from below and tried to grab his hand but we couldn't quite reach each other. My legs had been badly bruised in my earlier falls and were weak and not responding very well. I realized that if I couldn't even reach him from downhill and was having serious trouble walking myself, that I had no chance of carrying him. At that point I had to leave him and just hope that he could ride out the blockfall [fall of volcanic rocks] there and be rescued." He heard Williams crying for help as he made his way down the cone.

Macfarlane's head was spinning. The injury to his skull started to swell. His mind flashed to Dr. Peter Baxter's talk at the workshop the day before.

Fifty percent, Macfarlane thought. Baxter had said that when people are caught in a volcanic explosion, 50 percent usually survive. He winced as rock fragments battered his shins and thighs. *Fifty percent.* "It was a source of great encouragement for me," he recalled. Aloud, he repeated to himself as he ran: "We won't all die. We won't all die. We won't all die."

Everything was happening in slow motion. He ran and fell, his legs giving out as he tripped over the loose rocks. He got up and fell again. This time, his outstretched hands were scorched by the rocks. He knew that if he made it through the initial explosion without getting hit in the head, he had a chance to live. "I thought to myself that the burns didn't matter so long as I was still alive, Macfarlane recalled. "They could heal, and I could still crawl out even without hands."

Macfarlane expected to smell gas from the eruption cloud, but the plume didn't appear to be coming in his direction. Instead, it was moving north. *Thank God.* He had lost his gas mask in an earlier fall.

He continued downhill and found Mike Conway and Luis LeMarie huddled together in a shallow hole. The worst of the deluge was over, the largest rocks had fallen out, but the volcano continued to roar and burst with smaller explosions, heaving an occasional boulder 1,000 feet from the crater. Macfarlane crawled in with Conway and LeMarie, panting and unable to speak.

Conway shouted over the blasts for the other two to put their packs over their heads.

Macfarlane did as he was told.

Like a journalist in a war zone, Conway tensely reported to Macfarlane and LeMarie what he thought had happened to the volcano.

"It sounded like a pressure eruption. Like the volcano blew out the rocks in its vent."

The other two watched wide-eyed, the black cloud billowing in sinuous swarms.

"It *could* just be one blast," Conway said. "We may have a chance."

Macfarlane and LeMarie were silent.

"But if another eruption is coming, this time with a pyroclastic flow, we have to get out of here and get as far up the caldera wall as possible."

A pyroclastic flow would come with a lethal cloud of ash and steam that would incinerate them, suffocate them, turn their lungs to stone.

The caldera filled with clouds, which quickly condensed to rain. The three injured scientists remained close together. Macfarlane began to shiver in the cold rain. In a way, they felt an incredible sense of relief. The fusillade was waning, and they had survived the first eruption. But they knew there could be an even more vicious eruption coming. All three stayed very still, waiting like soldiers in a foxhole, listening for another deadly roar.

After five minutes, the rock shower was over, though ash continued to fall with light rain. On the summit, Alfredo Roldán pushed the newsmen off him and got up to look down into the crater. The clouds were pouring back in from the west, but he could still see the mushrooming eruption plume rising more than a mile into the sky, the wind distorting it to the north.

Roldán stared into the mile-wide caldera of the volcano. He stood in awe, mesmerized by the power of the volcano and its savage beauty, and he wondered if anyone else had survived.

CHAPTER 12: THE RESCUE

THE EXPLOSION SHOOK the surrounding countryside. Residents of Pasto and surrounding barrios came out to see the mountain, but the peak remained covered in clouds.

On the slope of the volcano, 1,000 feet down from the summit of Galeras, Marta Calvache's heart stopped. The seven people she was leading on the field trip stood frozen in disbelief. The initial explosion was loud, extremely loud, amplified by the walls of the valley where they stood; it seemed to come from everywhere at once. The first blast faded, but the mountain continued to rumble. A Canadian graduate student let out a scream and started to cry. "Some started to run in one direction and others in the other direction," says Calvache, "so I called the people going in the wrong direction away from the road."

Still crying, the graduate student turned to Marta. "What is it? What's happening?" To Calvache it was obvious. Galeras was erupting. "It was so clear that it was the volcano, but you try to think it's something other than the volcano," she says. She wanted to deceive herself, to draw another conclusion.

Patty Mothes also looked up toward the summit, trying to convince herself that it was a plane, thunder, blasting. Maybe it was military maneuvers or something going on at the airport to the north. But it was so loud, Mothes thought, louder than anything she'd ever heard at the eruptions of Ecuadorian volcanoes that she and her husband, Pete Hall, had witnessed numerous times. But in less than a minute, pea-sized pumice was raining down from the sky. There was no doubt. She thought about Hall. She knew he was on the other side of Galeras and she prayed he wasn't in danger. Mothes looked for a safe place for the group to hide and yelled at them to get behind a barrier of large boulders.

The graduate student cried harder, as most of the group covered their heads and huddled for several minutes, listening to the crackling of small rocks hitting the ground, feeling them bounce off their heads and packs.

Setsuya Nakada and Tad Ui didn't run, nor did they follow Mothes's direction to take cover. "Tad Ui started taking notes. I wanted to take a picture, so I got my camera from my jacket, but the mountain was in clouds," Nakada says. He looked at Ui and then at the frightened people crouched behind the rocky barricade, pebbles showering all around. After learning about Galeras for the past three days, the two knew they were standing on the safest side of the volcano. But for the Japanese scientists who had recently seen the wrath of Unzen firsthand, it wasn't hard to imagine what had happened at the top of Galeras.

After several minutes, Mothes decided the group was safe, for the time being at least. She tried to comfort the graduate student.

Mothes remembers: "I told her, 'Look, you're not soup yet. After all, it may not come this way. We're on the outside of the volcano, and it could very well go the other way.' "

Still huddled behind the boulders with the others, Mothes and Calvache looked at each other, then silently sized up the frightened group gathered behind the boulders. They knew they had to do something. The trucks were a good half-mile and 500 vertical feet above on the mountainside. The Canadian graduate student was now hysterical. Several others were visibly shaken. Most were under-prepared for hiking—they would be little help in a rescue. Heading for the vehicle would take them back up the erupting mountain, something that Calvache and Mothes were pretty sure no one in the group wanted to do. Without a radio to contact the observatory or reach someone at the top of Galeras, Calvache had no idea what might happen next.

She also knew that there could be a larger eruption on the way. In the past, Galeras's eruptions had lasted no more than a few minutes. But Calvache's experience constituted only an infinitesimal fraction of Galeras's million-year lifetime, and she knew that many stratovolcanoes started with a small eruption and then followed with a pyroclastic flow. If that were the case, they might all soon be dead. "So basically we just sent the people back," says Mothes. "We just said, 'Go back. Go down the road.' Because we didn't know if something else would happen."

Calvache looked at her watch. It was 1:45 P.M. Much later than a trip into the crater should last. "Normally, when you go to the crater, you go very early and you leave before noon. So I thought they were already out and going down to Pasto," Calvache says.

Besides, just the night before, the group of meeting organizers had emphasized that the expedition into the crater *must* be very quick—just in and out. Surely Williams must had gotten them out by now, safely on the caldera rim, she told herself. Deep down, though, she feared that this might not be the case.

Calvache made sure the others were on their way back down the trail to Pasto with Maria Luisa Monsalve. Then, as the mountain exploded three more times like a hideous bombing attack, Calvache and Mothes took off running up the steep trail toward the waiting trucks.

Back in town, at the Pasto observatory, all hell was breaking loose. The seismograph needles went crazy. A student monitoring the equipment who'd been expecting a tranquil day on Galeras was so startled, he nearly fell from his chair. It was 1:50, still lunchtime, and none of the seismologists had returned.

Instinctively, he grabbed the radio from the desk next to him.

"José! Come in! Come in!"

No answer.

He ran to the door of the observatory and looked toward the volcano. Nothing but clouds. Walking toward him in single file on the narrow sidewalk were Bruno Martinelli and Fernando Gil.

"Something's happening on Galeras!" he shouted to the two men approaching.

Martinelli and Gil, who hadn't heard the eruption as they were returning from lunch, ran past the student through the door of the observatory and into the front room to look at the seismographs. The seismograph needles were vibrating and scratching horribly jagged signals.

Martinelli quickly inspected the monitoring equipment and recognized instantly that Galeras had erupted. Yet he wasn't especially worried. It was nearly 2 P.M., and he felt sure the people on the mountain were well out of the crater.

Fernando Gil grabbed the radio and tried to call José Arles Zapata. Still no answer. The seismograph needles continued to draw violent waves while Martinelli tried again to raise Zapata on the radio, and Gil ran outside to see if he could see the volcano. Two minutes later, Fernando Muñoz came running into the observatory, his face flushed. People on the street were talking about the eruption; soon the small room that held the six Galeras seismographs was flooded with people. Everyone who worked at the observatory came to see what had happened, to see if there was any news from the top.

In the chaos, Ricardo Villota headed for the observatory truck. Villota knew that two of his good friends, José Arles Zapata and Carlos Estrada, were on the mountain, and he was terrified to think what might have happened.

"I told Fernando, 'I'm taking the Jeep and going back to the mountain,'" Villota says. "And he told me. 'No! No way. We have no idea what's going to happen.'"

Villota ignored Muñoz's warning, ran back outside, and jumped into the observatory's Jeep. At the same time, Milton Ordoñez, the young scientist who had been working with Geoff Brown on gravity measurements the day before, came riding up to the observatory on his bicycle.

"Ricardo yelled at me, 'Galeras erupted! Our friends may be hurt. We've got to see what we can do,' " Ordoñez says. He jumped into the observatory vehicle with Villota and they raced back up to the volcano.

Fernando Muñoz instantly took control of the observatory. He shouted commands at Adriana Ortega, who had just returned from eating lunch at her home nearby. He barked commands at a group of students who were huddled in a corner, worriedly speculating about the fate of the group on top of Galeras.

"Fernando told us, 'We need strong people here. We need only those people who are going to work. If you're not working, then go home,' " says Ortega.

The phones started to ring. As the radio reported that dozens had been killed in the eruption, family members, reporters, and foreign consulates called, pleading for information. No one knew exactly who was in the crater, not even how many had gone into the volcano. The park service had a list of the seventy people who had taken out permits, but there was no final list of the people that had descended with Stanley Williams that morning.

One hundred and eighty degrees around the circumference of Galeras, below the volcano's western slope in the Azufrado Valley, Hector Cepeda's group was still trying to figure out what the source of the explosive sound was. Clouds blocked the view of Galeras, and the sound ricocheted across the wide, misty valley and made it impossible to be sure of the direction of its origin. Some of the scientists gathered high on the valley slope, out of the way of a possible pyroclastic flow. But a few still weren't convinced it was the volcano they had heard, even John Stix, who had been to Galeras a dozen times over the last several years. "I remember asking some local people if there was blasting going on in the area, or some kind of road construction. So we really didn't know what to make of it at all," Stix says.

The weathered campesino who had been asked the question shook his head, amazed at the ignorance of these men who called themselves scientists.

"No, señor, el volcán esta despertando." Galeras was indeed wide awake.

Stix had planned to go into the crater today. He was not sure what had happened up top, but he remembered the July 16, 1992, eruption. Then, boulders as big as refrigerators were heaved from the mouth of the volcano when the lava dome blew into a million pieces of jagged, incandescent bombs. He was extremely relieved he wasn't there right now.

Those who had scrambled up onto the ridge above the Azufrado Valley included Fraser Goff and Gary McMurtry. They were not confused about the sound: it was the sound of a volcano erupting, and they wanted to get the hell out of there. Goff knew that his colleagues Andy Adams and Alfredo Roldán were wearing hard hats, gloves, and full-face gas masks, and he hoped that these Los Alamos safety measures had been enough.

From behind the mountain, radio transmission to the Pasto observatory was impossible, so Hector Cepeda took the bus and went to the police station in nearby Consacá to call the observatory while the others waited on the ridge above the Azufrado Valley. Thirty minutes later he returned. In the same calm manner he had used as their field trip guide, Cepeda told the group that the observatory was reporting that a much larger eruption could be on the way.

In the gray drizzle, the group hurriedly climbed back into the old school bus. Light ash started to fall from the sky. The mood was grave. No one wanted to speculate about the fate of their friends and colleagues.

For Gloria Patricia Cortés, a geologist from the Manizales observatory, it was a particularly grim moment. On the rainy night of November 13, 1985, Cortés was at home in Manizales, where she lived with her parents. Then a geology student at the University of Caldas, the usually affable Cortés was irritable. She had taken a makeup exam at the university that afternoon and had been forced to miss the field trip with her paleontology class bound for Ibagué. But her friends hadn't made it to Ibagué that night. They were in Armero. By the next morning, half of her classmates were dead.

Cortés was devastated. Wracked by guilt for her own good fortune, she had quit going to the university. After several months, Cortés's father, tired of seeing his daughter in so much pain, pulled her out of her bed—where she now spent hour upon hour, alone—and put his arm around her shoulder.

"My father told me the one thing I can do for my friends," Cortés says, "is to go and learn about the volcano. Learn about what happened to my friends in Armero, and help make sure this never happens again."

Cortés took his advice. Now, seven years later, she had planned to go to the crater with her good friend José Arles Zapata. But like so many others at the workshop, she had been told there wasn't room.

Pete Hall, who had been translating Hector Cepeda's field trip commentary into English for several of the non-Spanish speakers in the group, sat near the rear of the bus and looked out of the window. Hall was of course worried about the people on the field trip to the crater, but privately, he knew he'd been blessed. His wife, Patty Mothes, had decided at the last minute not to go to the crater. He felt sure she was out of danger.

Inside the caldera, fifteen minutes after the eruption, the explosions had subsided and the volley of rocks was over. Mike Conway, Luis LeMarie, and Andrew Macfarlane wanted to take advantage of what they suspected was a temporary calm to get out of the caldera as quickly as they possibly could. The volcano roared, giving off a constant, shrill hiss as it spewed black ash and steam. The tumultuous cloud had become a fiery orange at its base, and it grew thicker and blacker by the second.

Conway saw it first and pointed to the crater. The glowing incandescent ash had to be a minimum of 1000 degrees Celsius, an ominous sign. Conway thought it could mean that a pyroclastic flow was on the way. They needed to get out of the volcano, he told the others. But first they would have to get off the cone into the lowest part of the caldera, the most dangerous place to be if a pyroclastic flow, driven by gravity, poured from the crater.

The three stood unsteadily, badly burned, their hands and faces bloodied, and began to climb down the cone and head toward the caldera wall. The wound to Macfarlane's skull had left him dizzy, and he walked awkwardly down the slope, his head throbbing. Blood trickled down from his forehead into his left eye and dried, making it impossible to open.

The trail was grueling, and their pace was slow. Behind them, Galeras was still exploding—with an occasional crack that sent a volley of boulders into the air. Macfarlane held on to his pack as tightly as he could.

The faster they tried to move, the worse time they made. The three men would take a few steps, trip, and fall on the rubble-strewn ground, burning their outstretched hands and arms on rocks that still held the tremendous heat of the explosion. Conway, who had been injured the least by the barrage from the eruption, walked more quickly and shouted encouragement to LeMarie and Macfarlane. He was terrified that the two were moving too slowly to escape another eruption.

Macfarlane was exhausted. His head was spinning, pounding. Things seemed to move in slow motion as he walked. He stumbled and fell on hot rocks, got up and fell again. He was so incredibly fatigued that the only thing that motivated him to get up when he fell was the pain of the scorching rocks as they burned through his clothes and into his skin. "Between efforts to walk, I collapsed onto my back and watched the ash column, which continued to be incandescent in its lower part. I had to be careful where I dropped, because there were fragments of hot rock up to football size all over the ground on the caldera wall, and if I fell in the wrong place I would land on one," Macfarlane recalls. He forced himself to keep moving.

*   *   *

It took twenty minutes for Marta Calvache and Patty Mothes to hike up to the truck. The driver was waiting for them. Calvache grabbed the radio from him.

"Come in, Pasto!" she called.

The radio instantly crackled. It was Fernando Muñoz at the observatory.

"Marta!" he yelled. "Get off the mountain! There's been an eruption and it looks like a bigger one is on the way!"

"I'm going up," she said, firmly. "Somebody may need help."

"I said, 'Are you crazy?' " says Muñoz. " 'The volcano is so active! Don't go inside the crater. There are so many earthquakes happening. This is horrible.' "

Calvache never answered. She and Mothes climbed into the truck with the driver and headed for the summit.

The two women didn't speak as the driver careened around steep switchbacks. Coming around one sharp turn, they nearly ran head-on into a truck full of military police racing down the mountain. Calvache's driver slammed on the truck's brakes and the truck slid sideways toward the cliff. Both vehicles stopped dead and were in a standoff on the one-lane dirt road. Mothes jumped out of the Landcruiser and went to the other vehicle. The driver was shaken. Sweat crept down his deeply creased forehead. In the back of the pickup truck, Mothes saw that one of the policemen had a cloth wrapped around his forearm, saturated with blood.

The man in the passenger side leaned over so he could see Mothes. "He was a black man and you could almost say he was white with fear," Mothes says. "He said one of his workers had been working around the installation, and one of the blocks had fallen and cut his hand off. So they were taking him down to the hospital."

Mothes thought that the man who was talking to her must be in charge and could be of some use to them at the summit. "Come with us," she commanded. "We need your help."

The sergeant looked horrified and explained again to Mothes

that he needed to get the other officer to the hospital. Mothes would not relent. She told the driver of the military truck to take the injured man to the Pasto hospital, then yelled through the front seat of the truck to the sergeant, "You're coming with us."

At the summit of Galeras, thirty minutes after the eruption, the newsmen had regained their composure. They asked Alfredo Roldán if he thought it was safe to hike the 100 yards to the police station—they wanted to take cover in the concrete hut in case the mountain blew again. Roldán said yes, and the three of them walked carefully around the hot stones. The ashy rainfall was beginning to subside.

Roldán heard the reporter get through to Civil Defense and ask for emergency assistance, and he heard Civil Defense return the call. But Roldán was still not happy. Where he came from, the Guatemalan Civil Defense were a bunch of worthless *chafas*. He was sure it was the same here in Colombia; they'd do more harm than good.

He ran outside to the front of the police station, facing Galeras's crater. The cement sidewalk was covered with blood, and there was a glistening red-brown trail leading east from the station to the dirt area where the cars were parked. Roldán left the newsmen and hurried back to the caldera rim where the rope was fixed and waited, pacing. *Please, God, let them be alive.*

After several minutes, Roldán saw the rope begin to move and he knew that someone must be coming. Slowly, figures began to emerge from the cloudy caldera. Fabio García and Carlos Estrada came out first, moving ghostlike over the rim. They were dirty with ash, but Roldán could see that they were okay.

García took off running for the police station, and Roldán grabbed him. "I asked him, 'What happened to Andy?' " García told Roldán he didn't know. He hadn't been able to see Adams from where he was in the volcano.

Together they ran to the police station, where the journalist immediately tried to interview the two Colombians. García pushed the small reporter aside, and he and Estrada went to the radio, unaware that the others had already called for help.

"The radio would only call to Civil Defense in Popayán. I tried to call the observatory, but I couldn't. I told Civil Defense to send an ambulance, and they told me one was coming," he says.

Fabio García grabbed the microphone. "I told them to send a helicopter," García says. "They told us the helicopter didn't have any gas."

Carlos Estrada paced back and forth in the small room, worried about the others in the crater, especially his good friend José Arles Zapata. He told Fabio García that he was going back down into the volcano, and he wanted some help. García had no intention of going back to the crater. He was still incredibly shaken. He told Estrada to wait, that the Red Cross and Civil Defense were on their way. But there was no way Estrada could wait, and he ran back down the ridge to the rope.

Alfredo Roldán left the police station, again trying to avoid the pestering reporters. A Civil Defense convoy had arrived with a truck and two ambulances. There were now a dozen men at the summit, yelling orders and running back and forth from the police station to the caldera entrance, but none of them would go down into the volcano. Roldán couldn't blame them. He had no desire to die a horrible death either. He dug the heel of his boot into the dirt. Andy Adams should have been right behind García and Estrada, but he still hadn't made it out of the crater. Roldán felt bad for leaving him behind, for being irritated with Adams's slow pace, but what could he do? He couldn't carry anyone out if they were injured. The slope was hard enough for one person alone. Roldán went to the crater rim, to the top of the fixed rope, and looked down. The crater was once again filled with thick fog and he couldn't see anything. He called down to Adams, but there was no answer.

Hovering behind a small boulder at the base of the caldera wall, Andy Adams said a rare prayer as he watched the wind carry the mushrooming cloud away to the north. The perspiration on his body instantly felt cold. "My hard hat had saved me from the rocks, and now the eruption plume was going in the opposite direction. So, for the first time, I felt like I'd make it out alive," he recalled. His breath came more easily, but he started to feel sharp pains around the back of his neck. He removed a glove and felt above the collar of his padded suit. His neck was covered with burns that were starting to blister.

It had been thirty minutes since the eruption, and Adams finally conjured up the nerve to attempt to get out. He was terrified that another explosion would happen while he was exposed on the caldera wall. Going up the rope was even more difficult than he expected. It took fifteen minutes, but somehow he managed to pull himself to the top of the rim.

After what seemed like hours, Alfredo Roldán finally saw Adams's bulky form emerge from behind the caldera wall. Adams barely took two steps before the reporter shoved a microphone in his face, firing off questions rapidly in Spanish and demanding that Roldán translate them into English. The video shows Adams start to answer a question via Roldán's translation, but stop midsentence.

"I'm not doing this," he yelled, thrusting a dirty, trembling hand over the microphone, and pushing the journalist away. All Adams knew was that he wanted out of this nightmare, and the sooner the better.

The cameraman was capturing it all on video. It should have been a grand day for him. It was the scoop of a lifetime. He knew it was good journalism. Three days' worth of the world's top volcanologists looking into his camera and telling them that the volcano was "muy tranquilo." And then Galeras had erupted, right when these same scientists were in the mouth of the volcano.

But he wasn't thinking about how successful his documentary would be. He could only think of one thing. Like Gloria Patricia Cortés, he had survived Nevado del Ruiz. He had gone to visit a friend in a nearby city that night. In the morning, his mother and father, his brother's wife and their three children, and hundreds of his friends and neighbors were buried under millions of tons of wet earth. And the city where he was born was permanently erased from the map.

As he followed the manic reporter with his camera, he made a promise to himself and to God: If he lived through this day, he would never come back to Galeras, or any other volcano, ever again.

As Mike Conway, Luis LeMarie, and Andrew Macfarlane stumbled down the cone and across the caldera moat, clouds came and went, and Galeras turned from clear to cloudy and back again in a matter of minutes. Finally, after nearly an hour, they reached the base of the amphitheater wall. "When we got to the caldera wall and started up, I dropped my pack because it seemed that the rocks were not falling anymore, and that even that much weight was too much to lug up the slope," Macfarlane recalled. He was so completely drained of strength that he sat down there on the rocky slope. The clouds had once again cleared, and he could see all the way to the crater. He stared at the eruption column, pouring out ferociously, thick and black. The situation seemed more and more ominous by the minute.

LeMarie was several yards ahead of Macfarlane and moving slowly up the slope, his legs collapsing beneath him every few steps. Time after time he fell, struggled to his feet, stumbled, fell again—his spirit flagging each time he landed on the burning ground. Finally, he called ahead to Conway, telling him to go on ahead, to get out of the volcano and try to get help for Andrew and him. Conway nodded and kept moving doggedly toward the climbing rope. LeMarie rested for several minutes and thought about the others back at the crater, wishing he were strong enough to go and help them. He could barely move, but eventually forced himself back to his feet. "I continued alone and tried to climb the internal wall of the crater. That was the most difficult part. Thinking about my family gave me the spirit to continue," recalls LeMarie.

Andrew Macfarlane also summoned the energy to stand, though his legs were burning and felt like they would snap beneath him. He staggered toward the caldera wall, moving just inches at a time, his head cloudy and throbbing, aiming unsteadily for the orange-painted boulders that marked the way to the rope. He tried not to consider the fact that even if he reached the rope, he wouldn't have the strength to pull himself out. He could no longer see Conway or LeMarie and hoped they had made it out, that they were getting help. He kept moving toward the painted boulders, thinking that the closer he was to the trail, the sooner he would be found.

Macfarlane tripped and fell again, his right hand landing on a hot rock, scorching his flesh. He lay on the jagged, burning rubble, shivering as the rain and cold air chilled him and hypothermia started to set in. "I was calling for help and I was delighted to hear Stan calling for help from across the moat," Macfarlane recalled. Somewhere behind him, Stanley Williams was still alive.

When Mike Conway pulled himself to the top of the rope and reached the summit, Carlos Estrada was waiting to grab him. The video shows Conway's face covered with blood. His yellow Gore-Tex jacket was full of black-rimmed burn holes, his knit cap was stuck to his hair with dried blood, and his face was black with dirt.

Estrada led him up the ridge to the police station. Red Cross had now arrived and there were swarms of uniformed people on the ridge, all with anxious expressions. Conway, initially grateful to be alive, became furious.

"Why isn't anyone doing anything?" he asked. It had been over an hour since the eruption. Not one of these guys had gone down to help? He stumbled into the police station, where he found Andy Adams and Alfredo Roldán. He yelled at Adams, his voice high and angry. "What the hell are you doing? We need help right now!"

The videotape shows the exchange between Conway and Adams: "Listen," Conway says, "Andy Macfarlane and Luis LeMarie are hurt. They can get back up here, but they need help." He waits a moment, but gets no response from Adams. "Someone needs to go down there and get them out!" Conway snaps.

Andy Adams looks away, defeated. "Well, then, we need to send someone." But it sure as hell wasn't going to be him. In fact, besides Carlos Estrada, there wasn't a single person willing to go back in.

Seconds later, Ricardo Villota and Milton Ordoñez arrived at the summit; Calvache and Mothes's truck arrived right behind them. The vehicles in the lot were covered with rock and ash, and one of the Jeeps had two broken windows. Smoking rubble littered the crater rim; a white-hot rock sizzled when Mothes spit on it. Calvache leapt from the Landcruiser and followed Villota and Ordoñez along the ridge, past the milling rescue teams, to the fixed rope. Without hesitating, Calvache started down the slope on the heels of Ordoñez and Villota.

Mothes jumped from the truck and yelled at the sergeant. "I want you to radio for help right now."

"We called already—when Galeras exploded."

"Call again! Tell them we need stretchers and ropes and reinforcements, and we need a helicopter, and we need it *right now*!"

In the next room she found Alfredo Roldán, Andy Adams, and Mike Conway helping the battered Luis LeMarie into the building. Conway looked bad, but LeMarie, Mothes's good friend and colleague, was terribly injured. He was hunched over, supported on both sides by Adams and Roldán. LeMarie's clothes were torn and bloody, his hands gnarled and covered with ash and dried blood. Mothes crouched down to look at LeMarie's face.

"Lucho?"

LeMarie stared at Mothes, his eyes half shut. "*Hola,* Patricia," he said to her, his voice low and rough.

"Are you okay?" she asked.

"My legs hurt very bad. I think they're broken."

Adams and Conway explained to Mothes how LeMarie, with two broken legs, had somehow managed to get halfway up the steep slope under his own power. Then Carlos Estrada, the driver, had gone down the wall and pulled him out. They had all been in the caldera and seen the eruption. Several people were dead.

"Andy Macfarlane is at the bottom of the caldera wall," Adams added. "I seriously doubt that anyone else survived."

"I'm going down," Mothes said.

Adams remembers: "I handed her my hard hat and told her to put it on, that it saved my life down there." Then he and the other men at the police station watched as Mothes ran down toward the erupting volcano.

Roldán, Adams, and Conway carried Luis LeMarie to a cot in the police station and helped him lie down. Roldán searched the station for medical supplies. "I checked in the first-aid kit hanging on the wall, and there was some paint, PVC glue, and a hangover remedy," he says. Roldán looked through the open door to see if there was mud on the ground from the rain. He knew that mud was excellent to put on wounds. But the rain that fell on the barren soil evaporated immediately because of the heated surface.

The reporter ran over to the cot where LeMarie lay and held his microphone to the scientist's bloodied face. "Tell us what happened," he sputtered. "What happened to everyone down there?"

Fabio García tried to push him away, but the reporter persisted. His voice barely audible, with his bloodied hand outstretched, moving up and down as if to help him keep count, LeMarie listed those who were with him near the crater when Galeras erupted.

"Mike, Andy Macfarlane, Stanley . . . José," he said slowly, choking on his words. "José is dead. José Arles Zapata from Pasto. He was hit in the head with a rock."

García was listening to LeMarie speak. He had already heard someone say that Zapata was dead, but he had wanted to hear from another source. As the only senior scientist for the National

Institute of Geology and Mines on the site, he knew that now he have to give the observatory the horrible news.

Patty Mothes ran through the gathering of Red Cross and Civil Defense people. Ninety minutes had passed since the eruption and none of the rescue workers had dared descend into the caldera to look for survivors.

Mothes yelled, "We need help! Don't just stand there—somebody needs to get down there. And bring a stretcher and blankets."

She ran to the rope, just minutes behind Calvache. The caldera of Galeras was still full of clouds. "I heard Stanley yelling, 'Help me, help me,'" Mothes says. "I called back, 'Stan, we're coming.' It was totally foggy, and the fog kind of attenuates the sound waves, but I could hear him, so I knew he was alive, and that gave me all the more reason to think, 'Okay, I'm going down.'"

Ricardo Villota and Milton Ordoñez were near the bottom of the rope—with Marta Calvache halfway down—when Mothes arrived at the top. Villota and Ordoñez made it down and hiked toward Macfarlane, who was still lying on his back facing the crater, yelling as the two Colombians dislodged rocks from the slope.

Macfarlane wasn't moving out of the way, so Mothes and Calvache moved 10 yards farther east along the ridge to descend, still trying to avoid kicking rocks onto the men below. Without a rope, the steep slope was extremely dangerous, and the two women made slow progress.

Villota's field boots allowed him to move much faster than Ordoñez, who wore flat-soled shoes, and Villota quickly reached Andrew Macfarlane. The red-haired *gringo* looked horrible. He was trembling in the cold, covered with bloody cuts, and completely soaked with perspiration. Ordoñez arrived moments later and began tending to him.

"I could tell Ricardo really wanted to go and look for José Arles, so I told him to go ahead. I would stay with Andrew Macfarlane and wait for Dr. Marta to come," Ordoñez says.

Villota disappeared into the mist.

Mothes and Calvache reached Macfarlane moments later. Mothes took off her scarf and wrapped it around his head wound.

"I told Marta that I'd heard Stan calling for help, that he must be alive, so Marta took off to find him, and I stayed with Andy Macfarlane," Mothes says.

"You're going to be all right," Mothes told Macfarlane encouragingly. "Help is coming and we're going to get you out of here." Her immediate worry was that Macfarlane would go into shock. She rubbed his arms and legs and told Ordoñez to do the same.

"Patty Mothes was talking to me and encouraging me and rubbing me to get my circulation going, and encouraging the others to do likewise," Macfarlane recalls. "She apparently thought I was John Stix and kept calling me 'John,' but I didn't care at that point. She was an absolute flesh-and-blood angel of mercy." He barely knew Mothes. He had only met her three days before, and now she was risking her life for him. He tried to open his mouth, to say thank you, to say anything, but he was shivering so much he was unable to speak.

From the base of the sheer wall, rocks blocked the view to the top of the caldera rim, but Mothes could hear more people arriving. When rocks began to fall around them, she knew someone must be descending the rope. "Be careful!" she shouted. "You're kicking rocks onto an injured man."

One of the park rangers had arrived with an aluminum stretcher and some nylon rope. Mothes and the two men lifted Macfarlane into the stretcher and fastened him in with canvas straps. The ranger told Mothes that Civil Defense and Red Cross workers at the summit were preparing to pull Macfarlane out. Soon, a steel cable was lowered to the base of the caldera wall from above. The rhythmic sound of a helicopter came and faded away. Mothes knew that it was too cloudy for the helicopter to land, and it would be an extremely dangerous landing, even on a clear day, since wind over the narrow, rocky summit came in high-velocity gusts.

Awkwardly, struggling for purchase on the steep, rocky slope, Mothes and the ranger and Ordoñez slowly hauled Macfarlane as far up as they possibly could, while a crew at the top gathered up the slack in a steel cable.

The team above began to pull, and Macfarlane found himself being dragged head-first up the caldera wall. The metal litter bounced as it was hauled up the slope, slamming Macfarlane into the caldera wall again and again. He screamed as his fractured skull repeatedly hit the back of the stretcher. The straps that Mothes and Ordoñez had used to secure him were loose, and his injured legs, unable to support his weight, collapsed as he slid to the bottom of the stretcher. The pain was so extreme that he nearly blacked out.

More than thirty people gathered around him when he reached the summit. He was quickly carried back toward the police station and loaded into a waiting ambulance.

Ricardo Villota had never before been afraid of Galeras. Now he was terrified. As he ran down the steep slope of the caldera and across the flat, rubble-strewn moat, the fog lifted, and he could see the turbulent clouds roaring from the crater. As he made his way across the volcano and came to Galeras's cone, his heart sank. He was still 100 yards away, but there was no mistaking the bright yellow jacket of his good friend José Arles Zapata, who was lying facedown with both arms stretched above his head. Villota stood frozen in the mist. He and Zapata had stood at this very spot just four weeks earlier, measuring a large depression in the volcano. They were laughing and joking. That day, unlike today, Zapata had worn a bright yellow hard hat.

Villota began to walk again, slowly—moving silently toward the body—hoping against hope that he would see his friend move, breathe, anything. Suddenly, from his left, he heard someone calling.

"Help me. Please, help me. . . ."

It was Stanley Williams. Villota glanced one more time toward

Zapata, then quickly changed course, ran to where Williams lay, and dropped to his knees.

"I said, 'Dr. Williams, are you okay?' and he said 'My leg . . . My leg is broken,' " says Villota.

Williams's face was bloody, his upper lip pulled back in a horrifying grimace. Villota looked at Williams's leg and winced. "Later, when someone took off his boot, the leg broke even more—in a bad way," says Villota. "But he was conscious because he said he recognized me and knew who I was."

Villota looked again toward Zapata, whom he could see lying facedown just 20 yards farther up the slope. He knew Galeras could blow at any second, but he didn't care. All he cared about was his friend. So when he saw Milton Ordoñez and Marta Calvache moving through the field of boulders, he decided to leave Williams. "Dr. Marta and Milton are coming," he quickly explained. Then he headed back up the slope of Galeras's cone.

Zapata wasn't moving. There were rocks that had burned through his clothes and skin and were embedded in his flesh, some all the way to his bone. His light-colored pants had caught fire and were ripped and blackened. From the top of Zapata's head to the nape of his neck, his skull was removed, exposing his gray, bloodied brain. Villota was stunned, unable to move. The injuries to Zapata's body were grotesque, nearly unbearable to look at, and yet Villota still could not comprehend: he had to see his face. He grabbed Zapata's right shoulder and turned the body over; dead eyes stared back blankly at nothing. Blood from the skull wound had dripped down his temples, but his handsome face was free of cuts or burns.

In a desperate attempt to revive the languid corpse, Villota lifted Zapata's body by the front of his yellow jacket. Milton Ordoñez saw what had happened from 20 yards away. "It was terrible," Ordoñez remembered. "José Arles had a cut in his brain and when Ricardo sat him up, the brain fell out on the rocks."

Villota screamed and let go of the front of Zapata's jacket. The body dropped with a thud.

"I yelled at him, 'Ricardo, he's dead. There's nothing you can do,' " Ordoñez says.

Villota was hysterical. He held his arms around himself and rocked back and forth, his head bowed to the ground. He began to sob. "I felt desperate when I found José Arles," Villota says. "I had worked so many times with him. We were working every week together, and we were friends, and I was devastated when I saw him dead."

Ordoñez ran to Villota, then grabbed him by the shoulder. Villota continued to sob uncontrollably until Ordoñez slapped him hard across his face. For several moments, the two men sat next to the body of their friend. Finally, Villota let Ordoñez convince him to help look for other survivors, and the two men got to their feet.

Nearby, lying in the smoldering rubble, was a piece of Zapata's gas mask; embedded in it was a rock from Galeras melted into the plastic filter. Villota took it, held it in his hand for a moment, and with tears streaming from his eyes, put it in his jacket pocket. Then he followed Ordoñez up the slope.

Only 10 yards away were three other bodies. Villota and Ordoñez had never seen them before. They were dressed like locals, and Ordoñez could tell they weren't part of the science workshop— just hikers visiting Galeras. Two of the bodies were obviously young teenage boys, and the third was a grown man. The man had been nearly decapitated; one of the teenagers was missing part of his skull. One of the boys had a hole clean through his leg where a golf ball–sized rock had gone in one side and out the other. The bones of all three were clearly exposed where burning rocks had melted their flesh.

There was nothing they could do here, so Ordoñez and Villota continued, looking for other survivors along the circumference of the cone, even as the crater continued to rumble and erupt black ash and steam. Ordoñez went east and Villota went west. To the east, Ordoñez didn't find any signs of victims or survivors, and he continued walking in an arc to the north, keeping an eye out for boulders to duck behind, gambling that Galeras would not erupt again.

To the west, Villota came upon another body. The body was in two pieces. Both the torso and lower limbs were completely cooked and blackened. He knelt down before the upper body and removed a ring from a charred finger. He turned the gold band so that he could see the inside: *26–6–1991 Alicia Maria*—the wedding day, a year and a half before, of the young college professor, Carlos Trujillo, and the name of his bride.

Villota felt himself slipping back into despair again. He took a deep breath, gasping on the tendrils of poisonous vapor that whisked around him.

When Ordoñez came upon him from the north, "Ricardo shouted to me, 'I found Carlos Trujillo. He's dead, too,' " Ordoñez says.

Ordoñez got close enough to see the charred remains of the professor. "Come on, Ricardo," called Ordoñez. He grabbed Villota's hand and pulled him to his feet. It seemed that there were no more survivors, so together they set off to help Marta Calvache with Stanley Williams.

Calvache had witnessed the heartbreaking scene of Ricardo Villota with José Arles Zapata from a distance. She knew that the bodies further west on the cone were dead, but to the east, she could see the crumpled form of Stanley Williams. Patty Mothes had heard him crying for help, so she knew he was alive. She finally made it to his side and knelt down.

"Are you okay, Stanley?" she asked.

"Help me. Please, get me out of here. I don't want to die."

Calvache saw blood trickling from behind his ear, sticking to his knit cap. There were burns on his face, and his clothes had obviously caught on fire.

"It's going to be okay, Stan. We're going to get you out of here."

Williams's leg was horribly broken and he was covered in burns, but he was coherent. Calvache was relieved to see Patty Mothes and the ranger coming toward them. Also coming to help were Ricardo Villota and Milton Ordoñez. Williams had been quiet and didn't appear to be in too much pain, but then the park ranger took the lower half of his shin and tried to bend it back in place.

Williams let out a hideous scream.

"I told the ranger, 'Look, he's not gonna die from that,' " Mothes says.

But the three others seemed determined to straighten the fractured bone. Ordoñez asked Calvache for her knife and cut part of the Styrofoam cooler that Andrew Macfarlane and Mike Conway had brought with them. He made splints for Williams's leg and tied it with the laces from Williams's boots.

The ranger had a blanket with him, and Mothes suggested that they carry Williams in the blanket and try to get him to the base of the caldera wall. That way, she said, when the stretcher came, they could get him out as soon as possible.

Ricardo Villota helped the other four lift Williams and put him into the blanket. Then, without saying a word to any of them, the young Colombian got up and walked away. In shock, his heart nearly broken, he simply had to get out of Galeras.

Calvache and Mothes, along with Ordoñez and the park ranger, started to haul Williams out of the crater. Each held a corner of the blanket that carried him. He seemed unbelievably heavy and they moved very slowly, stopping to rest every few steps. "It's rough to make the walk by yourself. With four of us carrying Stan, it took two hours to get him to the cliff," Calvache says.

Galeras continued to rumble, and Mothes looked uneasily toward the crater, wishing she had a radio to get in touch with the observatory to see what the seismographs were showing.

The quarter-mile trip to the base of the caldera wall took nearly two hours. The volcano continued to roar, but there were no other signs of an impending eruption. The weather stayed clear, and they could see a helicopter landing on the caldera rim.

Through gritted teeth, Williams repeatedly asked what was taking so long.

"We'll get you there soon, Stan," Mothes promised.

Finally, they got him to the base of the wall where a dozen people were waiting. A Red Cross medic moved the others out of his way and knelt down over Williams. First he removed the Styrofoam splint and yelled for someone to bring him an aluminum brace. Then he took off Williams's bloodied cap.

"Oh, my God," gasped Calvache. She put her hand to her mouth, thinking she would pass out.

Under his dark knit cap, Williams's skull was crushed. Above his left ear a bloody, gaping wound exposed his brain. They quickly moved him from the blanket to a metal stretcher and fixed the stretcher to a steel cable. The Red Cross medic had a radio and called to the top of the rim.

"Okay. He's ready. But go slow. He's hurt bad," he yelled.

As a dozen men pulled from above, the metal stretcher bounced against the rocky caldera wall. Williams let out a yell and continued to scream as they dragged him the 200 feet up to the summit.

The small military helicopter finally landed on the rim of the caldera, and Williams was carried to it. The pilot took off immediately, heading for the hospital in Pasto.

As soon as Patty Mothes and Marta Calvache saw that Williams was out of their hands, they were overcome by incredible fatigue. The reserve of adrenaline that had driven them up the mountain and into the crater during the eruption seemed to evaporate. But there could still be others, Calvache thought.

Mothes felt the same, and the two women hiked back toward the cone to continue looking for survivors. "I went out and tried to find if there were more bodies. The volcano seemed to be giving out more sounds, and I didn't have a radio, and I was always looking for rocks to hide behind. I didn't find anyone on the first pass, and I didn't want to be the next victim," she says.

By 6 o'clock in the evening, it was cold and raining inside of the caldera. Twilight was setting in, and Calvache and Mothes, discouraged and saddened that they had had no luck finding anyone else alive, began to hike out of the volcano.

After delivering Williams to the hospital, the helicopter returned to the rim, but there were no more survivors to evacuate, and Mothes and Calvache silently boarded the small craft. Calvache looked down into the smoldering volcano. From the time of her earliest memories, Galeras had always been like a friend to her, to her family, to her people. An incredible sadness washed over her.

CHAPTER 13: THE AFTERMATH

SEVENTY PERMITS HAD BEEN ISSUED that day for visits to Galeras, and no one had any idea how many people had been inside the volcano during the eruption. Local radio blared a yellow alert, announcing that Galeras was in a heightened state of activity, and calls from Civil Defense warned to expect the possibility of dozens injured. In the Pasto hospital emergency room, there were too few beds and too few doctors and nurses. White-clad orderlies scurried along the hallways pushing wobbly gurneys into the hospital's receiving area to intercept the wounded.

At 3 P.M., three military policemen arrived at the hospital. They rushed into the emergency room, shouting for help and dragging the young officer whose hand had been amputated, his arm wrapped in a blood-soaked shirt. He shook as nurses unraveled his makeshift bandage; a tourniquet was quickly wrapped around his upper arm to slow the bleeding, and nurses began cleaning the wound while his two *compañeros* babbled maniacally about the explosion.

"There were a bunch of scientists down there when it happened. If they survived, there could be at least twenty more injured."

The older of the two, realizing that their *compañero* hadn't yet seen a doctor, yelled at one of the nurses. "He needs a doctor! Where is the doctor?"

A doctor arrived and began to treat the wound, yelling at the attending nurses for supplies. All off-duty personnel were called in, and the hospital quickly filled with anxious workers running in all directions, trying to find and prepare available beds and emergency supplies for the anticipated deluge.

After a long, painful ambulance ride down the mountain, Luis LeMarie arrived in the hospital's chaotic triage area. His face and hands were bloodied and black with ash. Fabio García had ridden in the ambulance with LeMarie, who was conscious but not speaking. García helped the Red Cross workers carry him into the hospital and lift him onto a narrow bed. Nurses swarmed around the bearded Ecuadorian and stripped him of his clothes, gasping at his mangled body. He didn't look like a victim of violence or a traffic accident—more like the mutilated prey of a horror movie monster. It was as if Galeras had chewed him up and spat him out. Lacerations like giant tooth marks gouged his arms and legs. He was covered with ghastly bruises, his stretched skin was crimson and raw, his limbs were swollen and bloated, looking ready to burst. His hands were covered with blisters and caked with dried blood and dirt.

Nurses quickly wheeled him into the emergency room where critically sick and injured were lined up in narrow beds along gray-white walls. The air was stale, and there was a loud clamor of voices and machines. Two stout women brought buckets of water, deep brown antiseptic, and sponges. They gave LeMarie several shots of anesthesia and then began to scrub his gaping sores.

Andrew Macfarlane had also suffered through a seemingly endless ride down the mountain. As the vehicle bounced along the rocky switchbacks, the adrenaline in his system waned, and the pain of his scorched skin became acute. The contusions on his legs swelled and throbbed, and he lay in the stretcher shivering with hypothermia, yet the pain itself was a source of relief: it was a sign that he was going to live.

Macfarlane reached the hospital an hour after LeMarie, and the medical staff readied to pull him from the vehicle. "I remember arriving in the ambulance and the doors swinging open at the hospital loading dock," he recalls. "There was a crowd gathered and they gasped in unison at my appearance, which seemed comic to me, even at the time." He was rushed inside and stripped of his wet clothing. He shook uncontrollably as the nurses began to scrub ash and gravel from deep gashes where rocks had cut through his flesh. There were thick, blistered burns on his swollen red hands. The rock to his head had given him a concussion, but the pain of his injuries kept him conscious, and he lay content at the mercy of the Colombian nurses.

It was just after 5 P.M. when the helicopter carrying Stanley Williams landed in a soccer field next to the hospital. Two Civil Defense workers brought him into the emergency receiving room just as Fabio García ran in. "I felt very bad for Stanley. He was unconscious, practically without clothes, and he looked horrible," García says. García thought he would be sick as he watched the nurses strip off Williams's tattered clothing, scorched skin peeling away with the tattered fabric. There were burns and blisters and gouges all over his naked flesh. His lower left shin remained attached to his body by only a few threads of bloody white fiber. The bone on the left leg was exposed and obviously fractured. The left side of his face was ripped away, his ear and skin carved from his head. In a small crater in his skull, slivers of bone were embedded in the gray matter beneath.

The physician on duty in the emergency room immediately called Pasto's only brain surgeon. He arrived quickly. Williams's condition was dire and the surgeon did not want to operate on Williams's skull without a CT scan that would show the internal injuries to his brain. But the hospital didn't have a CT scanner, and Williams would have to be driven across town to a medical lab. As more and more people filled the emergency room, anxious to see the volcano's victims, two nurses wheeled Williams through the narrow hallway. A television crew relentlessly filmed the chaotic scene. A microphone was shoved into Williams's face before an incensed nurse screamed at the journalist and batted it away.

As Williams was being taken to the waiting ambulance, the helicopter carrying Patty Mothes and Marta Calvache arrived at the adjacent soccer field. Mothes saw Williams being carted to the ambulance. "I went with Stan to get a CT scan somewhere across town. We were gone at least an hour and he kept talking: 'I'm so cold. Am I going to die? I want to be with my wife and kids. I don't want to die.' "

The CT scan revealed a subdural hematoma—blood had clotted between the layers of Williams's brain and fragments of skull were embedded in his brain tissue. When Williams and Mothes arrived back at the hospital, he was immediately taken to the intensive care unit and prepped for brain surgery.

The scene at the hospital had become even more frantic. The British doctor, Peter Baxter, was helping the physicians treat the wounded. Mike Conway, the least injured of those who had been on Galeras's cone at the time of the eruption, was treated for cuts and burns that covered his back, his head, and his hands.

Marta Calvache had called Bruno Martinelli, who had taken the seismology group back to his home to wait for news.

"Oh, thank God, Marta. I was so worried. Did you find anyone else?"

"No, I don't think anyone else made it. I'll call you later. Stan is really bad."

She hung up the phone and put her head in her hands. Calvache tried not to think about José Arles Zapata and Nestor García, and to focus instead on making sure that Williams got the best care possible, and that Arizona State University, which had been notified of the accident by Christopher Sanders, was doing everything they could to get Williams back to the United States. She thought of Williams's wife, Lynda, whom she had known ever since she started graduate school in Louisiana. Besides Williams, there were three other badly injured foreign scientists in the hospital and fifty others at the hotel that she was now responsible for. She knew she would soon have to give interviews to the press, and then the mayor and governor would be calling, perhaps even the president. *Stay strong,* she told herself. *Stay strong.*

After doing all they could at the hospital, Calvache and Patty Mothes left for Hotel Cuellar. "We went to the hotel to tell the other scientists what we knew, because at that moment we were not sure who was in the volcano and who was not," Calvache says. Most of the scientists were gathered in the hotel lobby, and there was complete confusion. The radio and television news reports had been terribly inconsistent, quoting dozens of injured victims. There were people reported dead who hadn't even been to the volcano that morning. Hotel Cuellar didn't have any phones in its guest rooms, so the scientists jammed the lobby and the hotel owner's crowded office, waiting impatiently for the phone to call their families.

Fraser Goff and Gary McMurtry had been at the hotel since midafternoon, when the school bus dropped off the scientists from their truncated field trip around Galeras. From the hotel roof, they had watched the military helicopter land and take off several times on the summit of Galeras, but they had no idea if Andy Adams or Alfredo Roldán had made it out alive. After a horrifying five-hour wait, at 7 P.M., filthy and exhausted, Andy Adams and Alfredo Roldán walked into Hotel Cuellar's lobby. "We just kind of hugged each other and cried for a while," Adams says. After a moment they pulled themselves together and began to recount the tale of the horrifying adventure. Goff quickly wrote a memo to Los Alamos stating that the entire team was okay and asking that their families be notified. He faxed the note as soon as he could get an open line.

It was after 8 P.M. when Marta Calvache and Patty Mothes arrived at the hotel and gave the crowd a debriefing as to what they knew: José Arles Zapata was dead, as were three other unidentified men, assumed to be local Colombians who had come to visit the crater. They had all been killed by falling rocks. Carlos Trujillo's body had also been found, but there were no signs of Geoff Brown or Fernando Cuenca, who had been on the southwest rim of the crater when the volcano blew; they were assumed dead. Nestor García and Igor Menyailov had been in the crater at the time of the eruption and were also assumed dead. Stanley Williams was in critical condition at the hospital and Andrew Macfarlane, Luis LeMarie, and Mike Conway were being treated for less critical injuries.

That night, exhausted and depressed, Fraser Goff and Gary McMurtry took Andy Adams and Alfredo Roldán to dinner at the hotel's second-floor restaurant. Goff presented the two with a bottle of red wine to which he had attached a tag that, written in a red marker, said: "red badge of courage." The usually boisterous Adams was pensive. He felt like he had been in a war. He would intermittently shake with chills and break out in a sweat, remembering the violence and his indescribable fear. He really wanted nothing more than to get the hell home to his wife and kids.

Goff held up his glass in a toast and the others joined him. Adams tried to keep his hand from shaking.

"To making it out alive."

"To surviving the volcano!"

"Hear, hear!"

Downstairs, Hector Mora, a scientist from the Manizales observatory who had worked with Geoff Brown's team the day before, was in the lobby of the hotel. He was helping the foreign scientists use the Colombian phone system when he received the dreaded call from the director of the National Institute of Geology and Mines. It would be his job to tell the families of the deceased what had happened. He set off with a heavy heart to the homes of the victims from Pasto. The widow of José Arles Zapata already knew her husband was dead, and he found her in a trancelike state, surrounded by family. There was little more he could tell them, so he offered his condolences and left. Delivering the news at the home of Carlos Trujillo was more difficult. Although his widow was somber and quiet, her brother was enraged. "We identified Carlos Trujillo by his gold ring, but his brother-in-law didn't want to believe that he was dead," Mora says. He was drunk and angry and yelling. He blamed the scientists for getting Carlos killed. I had to accept that, because I understood the situation."

Still, Mora was at a loss. It wasn't the fault of the scientists, it was the fault of the volcano. But Trujillo wasn't really a scientist, he was a teacher. He turned to Trujillo's widow, who sat stone-still on the living room sofa. "I am sorry, I am very sorry for your loss." He held out his hand, but he got no response, so he quietly excused himself.

Now he would have to go to the observatory and make phone calls to the families of Nestor García in Manizales and Fernando Cuenca in Bogotá. Mora was from Manizales and he knew Nestor García well. He thought about García's wife and kids, his parents and siblings. He held his hand over his eyes to try to hold back his tears. It was so horrible. And the others, what about the foreigners? He had no idea how to inform the relatives of Geoff Brown or Igor Menyailov.

At 9 o'clock, Fabio García returned from the hospital to Hotel Cuellar, accompanied by three Civil Defense officers. They would need pieces of the victims' clothing for the search-and-rescue dogs. They asked the hotel bellhop, Luis Ruales, a 45-year-old *Pastuso,* to collect an article of clothing from each of the missing victims. Ruales remembered instantly which rooms the foreigners were in, got the keys from the front desk, and went up the stairs. The Englishman was in 205, and the Russian was in 218. Ruales took a sweater from Brown's duffel bag.

Ruales had had a premonition that something bad would happen. "When the scientists came on Sunday, I helped them carry their bags." He had seen the equipment that the Russian intended to bring to Galeras—"metal things, frightening things"—and was convinced that harm would come to the visitors. "The scientists dropped something into the crater and made Galeras angry. That's why Galeras erupted," Ruales says.

This explanation for the eruption was held throughout Pasto— "Everyone knows it's true," says Ruales—and while the rest of Colombia and the media were talking of the harrowing tragedy, the

Pastusos were less generous. Much of the economic ruin and fear they had lived through since 1989 had been brought on by these very scientists and their catastrophic predictions. Now, after telling the people of Pasto to fear the volcano, they had ignored their own advice. The scientists had taunted Galeras, like a bullfighter taunts a bull, and the volcano had struck back. Galeras had never taken a victim in all of human history—until the day when the "experts" walked into its crater.

The yellow alert was laughed off by the *Pastusos,* who had been through it all many times before. It was a tiny, barely noticeable eruption, save for the fact that the scientists were right on top of the mountain when it blew. What little respect the locals may have had for the scientists was now completely gone. The governor and mayor openly scoffed at the volcanologists. *Diario del Sur,* Pasto's daily newspaper, published a cartoon poking fun at the ironic tragedy, with a giant boulder bouncing from the summit of Galeras and squashing a meeting full of volcanologists.

The media in general was on a rampage. Close-up shots of Macfarlane's and Williams's mangled faces plastered the front pages of national newspapers. "The Hellish Volcano," the headlines read. "Volcanologists Devoured by the Crater." "First the Roar and Then the Hell." Outside of Pasto, relatives were once again frightened for their loved ones who lived in the region, despite the *Pastusos'* reassuring phone calls that Galeras was just fine—as usual.

The frenzy at Hotel Cuellar had subsided somewhat by 10 P.M. Most of the scientists had reached their relatives and reported that they were alive and well. After giving an oral report on what he knew to a group of scientists, a fatigued Fabio García started up the stairs to his room. In the second-floor restaurant, a television tuned to the national news blared the eruption coverage. He listened to the roster of injured: "Stanley Williams de Los Estados Unidos, Andrew Macfarlane de los Estados Unidos, Fabio García de Bogotá." He was so tired, it almost didn't register.

Oh, hell.

He'd have to call his wife again and reassure her that he was all right. Weak from exhaustion, he walked back downstairs to the lobby and waited behind two other scientists who were making calls to their respective countries. He got through to his hysterical wife, whom he had already spoken to, and assured her that he was fine. She had heard the latest television report and was sobbing. "I told her, 'Don't worry. I'm not injured. The information is a lie,' " García says.

He handed the phone to the receptionist behind the counter and started up the wide stone stairs to the second floor. He walked to his door and turned the key, opening up his empty room. Next to the twin bed on the far wall was a canvas duffel bag, neatly zipped shut. Inside were the belongings of Fernando Cuenca, García's colleague at the National Institute of Geology and Mines in Bogotá. García stripped off his filthy clothes and walked into the shower. He stood beneath the warm water and ran his fingers through his thick black hair. His scalp was full of sandy pumice and gritty ash. The water ran black down his body, and he watched tiny pieces of Galeras swirl down the shower drain and disappear. García stayed in the shower until the water ran cold. Then he got out and dried himself off, shivering in the unheated room. He lay down on the twin bed, closed his eyes, and tried to sleep. Last night, he and Cuenca had been laughing and joking. Tonight, his *compañero* was dead.

Cuenca was a newlywed, and he had recently returned from Russia with a beautiful and sweet-natured Russian bride. García thought his heart would break. Before crawling into bed, he said a prayer for the young widow, who was now alone in Colombia— thousands of miles and a world away from her home. García remembers, "I could not get to sleep. When I closed my eyes, I had the permanent impression of the explosion, like a photograph, like a moving picture in front of my eyes, playing over and over and over. . . ."

Across town, the madness had not yet subsided at the Pasto observatory. Galeras was still rumbling, and Fernando Muñoz, sleeves rolled up, face gaunt, barked orders at a half-dozen scientists to analyze each quake and determine where it was coming from. The deeper the quake, the greater the possibility that another eruption was on the way—one driven by an escaping magma body and possibly accompanied by pyroclastic flows. There were hundreds of quakes, and the data came in from all six seismometers. While it was being analyzed, other scientists from both the Pasto and Manizales observatories helped answer the endless barrage of phone calls from the press and the foreign embassies.

But behind the hubbub of immediate concerns, what had happened earlier was eating away at Muñoz. What had gone wrong? Why hadn't the radio been working? Why had they been in the crater so late?

At midnight it began to rain, building from just a sprinkle to a deluge. As the rain pelted the street in front of the observatory, Muñoz and his colleagues grew even more somber, knowing that the temperature on top of Galeras would drop below freezing. "If anyone were still out there alive, it was not likely that they would survive the night," says Adriana Ortega.

At 4 in the morning, Milton Ordoñez arrived at the observatory to pick up a truck. He would go with the Civil Defense and Red Cross workers to resume the search. A caravan of ten trucks and ambulances negotiated the road up Galeras in the dark and reached the summit by 6 A.M. The top of Galeras was completely foggy again, and the thirty rescue workers stood on the summit, waiting for the sun to rise. All of them were well equipped with hard hats and safety gear, but none of them was relishing the idea of descending into the volcano that had just killed at least nine people. Three dogs in Red Cross vests—a black Labrador, a German shepherd, and a golden retriever—sat happily on the summit, receiving lavish attention from the anxious crowd.

After several hours, the clouds inside the caldera cleared, and the rescue workers slowly descended the rope into the volcano. They still didn't have a list of missing or any idea of how many bodies to look for. The crater of Galeras steamed slightly, the very picture of tranquillity. The teams spread out around the moat, making their way slowly toward the crater. The weather didn't stay clear for long, and walking around the rubble in the fog was treacherous. Milton Ordoñez was extremely nervous. He stepped carefully, always looking for boulders to dive behind in case the volcano decided to erupt.

The first order of business would be to remove the bodies of José Arles Zapata, Carlos Trujillo, and the three hikers. With the same method they used to pull out Stanley Williams and Andrew Macfarlane, they would haul out the dead, carrying them first on aluminum litters across the moat to the slope of the caldera wall, then strapping them in and hauling them up. Because of the steep angle, it took a dozen men at the top of the ridge to haul up each body.

José Arles Zapata's body was wrapped in a blanket and taken out first. The three hikers—according to news reports, a local high school teacher, his son, and the son's friend—were hauled out next. They had each died the same way, with a split skull, a rock penetrating the thin bone and soft brain tissue.

Finally, they strapped in the torso and legs of Carlos Trujillo and pulled the split cadaver slowly up the caldera wall. The work was gruesome and exhausting. The visibility was near zero most of the day, and the rescuers walked through fog, stumbling, often getting lost while they searched for bodies with the help of the three dogs.

Around the southwest rim of the crater, there were cooked pieces of flesh and smears of blood stuck to rocks. The workers collected the pieces by scraping them from the rocks into plastic bags, sure that the relatives of Geoff Brown, Fernando Cuenca, Igor Menyailov, and Nestor García would want something to bury. The mangled aluminum shell of Geoff Brown's gravity detector was recovered 1,000 feet from where Brown's team had been working. Also recovered were pieces of a gas mask, a glass vile full of gas collected by Luis LeMarie that, miraculously, had survived the eruption unbroken, and Andrew Macfarlane's shredded backpack. Because of the poor visibility, the search was exceedingly slow, and by 5:30 P.M., the exhausted Red Cross and Civil Defense workers began their hike out of the volcano, planning to resume their search the following day.

CHAPTER 14: CONFLICTING ACCOUNTS

BY THE MORNING OF FRIDAY, January 15, with five confirmed dead and a minimum of four bodies missing and presumed dead, the news hit the international media. In the United States, the sentiment within the scientific community was one of complete shock. Two scientists were confirmed dead, four were missing, and three hikers who had been near the crater had also been killed in the eruption. Stanley Williams had taken a rock in his head and was close to death in a Colombian hospital.

The scientists from the U.S. Geological Survey were incredulous. Stanley Williams had been felled by a rock to his head? "What happened to the safety procedures?" USGS volcanologist Dave Harlow wondered, echoing the sentiments of the geologic community. Others killed by falling rocks? Where were their hard hats? And why were tourists and reporters visiting the active crater while research was being conducted? The USGS scientists who had complained about the State Department's judgment for canceling their trip were now counting their blessings.

No U.S. scientist, however, was more shocked than Bernard Chouet. In 1992, Galeras had acted just as Chouet had predicted, sealing itself off and then showing the long-period seismic signals before it erupted on July 16. Had Galeras blown this time without any signs—without any signals at all? Chouet found this hard to believe. But if Galeras *had* shown signals that the volcano was unstable, why would Stanley Williams have taken a field trip into the crater?

The scientists at Hotel Cuellar were reeling in the aftermath of the tragedy. Trying to salvage the purpose of the gathering, they morosely worked to put together the report that they would submit to the United Nations. They talked in hushed tones about their dead colleagues, and those who hadn't made contact with relatives continued trying to get through on the lobby phone.

Tobias Fischer was in a state of shock. The young German graduate student had spent the last six months working at Galeras collecting gas samples for his master's thesis. Stanley Williams was both graduate adviser and father figure to Fischer; without Williams's encouragement, Fischer might not be doing the work he'd come to love. Williams's death would have devastated him personally. But at the same time, Williams had been hurt—and others had died—doing precisely the work that Williams had encouraged Fischer to do.

Fraser Goff was concerned about Fischer's safety and felt that Williams had been remiss about teaching Fischer adequate safety measures. "Our team adopted Toby," Goff says. "Stan's last orders to Toby were to go to the crater as often as possible, and we said, 'Don't do that. Work with us and stay alive instead.' "

A large group of scientists gathered that night at the hotel for dinner. Fraser Goff sat next to Meghan Morrissey, the Arizona State University graduate student who had been on the field trip with the seismologists. Morrissey had spent some time talking to Roberto Torres and Diego Gómez. She knew something about long-period seismicity because she had just begun to work with Bernard Chouet. After the eruption, the Colombian seismologists had told her that *tornillos* had appeared on the seismographs for two weeks prior to the eruption. When Goff began to talk about his team's plans to stay in Pasto and continue working on Galeras, Morrissey became worried that their lives would be in jeopardy. "At dinner, Meghan was talking about *tornillos,*" Goff remembers. "I told her I was planning on going back in the crater and she said 'Fraser, you've got to watch out for the screws.' " Morrissey told him about the science of long-period seismicity and to make sure that there were no *tornillos* before going back to the volcano.

Goff wasn't sure that Morrissey knew what she was talking about, but he planned on checking the seismicity every day before he and his team went to the crater. He would definitely look for the strange signals. The next morning, his team went to the observatory and asked to have a look at seismic data. He remembers that the Colombian seismologists brought out several of the seismographs from the days before the eruption.

One of the young Colombians pointed to a scribble in a seismograph dated January 13, 1993. "Now, after the eruption, Galeras isn't showing the *tornillos* that it was before, so it's probably safe to go back into the volcano," he said.

Goff couldn't believe it. There *had* been signals. Could it be possible that no one looked at the data? Or that they had simply kept it to themselves? All Stanley Williams and John Stix had ever said was that the volcano was completely quiet, not showing any signs of activity—even right up to the day of the eruption. Goff knew that to go into an active volcano without looking at the seismograph data was like playing Russian roulette. He had assumed when he sent Adams and Roldán into the volcano that the team leader—in this case Williams—had completely checked out all safety concerns. He hurriedly scribbled down the times and durations of the signals he saw on four different seismographs that were recorded just prior to the eruption. "In the back of our heads, we wondered what the hell was going on. But officially it would have been improper to say anything because it was already a very uptight situation, and we had another three weeks to work there."

Later the same night, Andy Adams sat down with the thick volume of meeting materials. Inside the binder were brief scientific papers on Galeras and the other volcanoes that had been discussed by many of the scientists at the meeting. Adams decided that he wanted to know more about the volcano that almost killed him, especially since it seemed like Goff was considering going back inside the caldera to take samples. As he thumbed through the first pages, the meeting schedule, the lists of participants, his colleague Gary McMurtry pointed until he came to a particular abstract. They weren't in the same order as the presentations. In fact, they didn't seem to be in any particular order. He began to read the very first entry. It was an abstract that had been written by John Stix just prior to the conference. Adams could not believe what it said.

State of the Volcano

The seismic and SO$_2$ data, taken together, suggest that either (1) the magmatic system has gradually sealed itself so that surface degassing is inhibited, or (2) the system is reaching the end of its life. The first hypothesis implies that gas pressure may be building beneath the dome, which will logically lead to explosive eruptions, while the second hypothesis suggests that the current eruptive cycle is ending.

Future Scenarios at Galeras

(1) It is quite possible that an eruption, similar to that of 16 July 1992, will occur in the next few weeks or months. The low seismicity and gas levels at the volcano resemble those before the explosion of July 16. This eruption does

not appear to have opened the conduit of the volcano to allow degassing to occur. Thus, the conduit may remain sealed; and gas pressure may continue to build beneath the edifice. This is clearly a serious scenario, and monitoring efforts should attempt to recognize any unusual changes such as the appearance of monochromatic long-period events or banded tremor, large daily fluctuations in SO_2 fluxes, and deformation of the edifice.

The fundamental question as of this moment is this: Do the current low levels of seismicity and degassing indicate continued unrest at Galeras, or do they simply reflect the waning stages of this particular period of activity? Several lines of evidence suggest that Galeras remains active and dangerous.

Adams thought he would be sick. Stix had predicted an explosion could occur in just a few weeks? His paper even talked about the need to recognize the long-period seismic signals—*tornillos*— that Adams and Goff had seen in the observatory this morning. None of this had ever been discussed at the meeting. Adams was furious. He, Alfredo Roldán, and everyone else who had been on that field trip had been told that the volcano was completely tranquil.

On the hotel television, Pasto news ran a video clip over and over: John Stix and Stanley Williams in the lobby of Hotel Cuellar proclaiming the tranquillity of Galeras. The TV producers couldn't get over the irony. Adams couldn't get over his anger. "I felt that Stan Williams had really screwed up," Adams says.

Fraser Goff's group, along with Tobias Fischer, planned to stay in Pasto for another three weeks, but the rest of the workshop participants left on Saturday. An air ambulance, sent by Arizona State University, was on the way to Bogotá. Surgery on Williams's brain had successfully removed a blood clot and fragments of skull. Both of his mangled legs were set in casts, and his burns were cleaned and treated. Williams's wife, Lynda, having been told that she might arrive to find her husband a vegetable, was also on her way to Bogotá. She joined Williams in Cali and the two were flown back to a hospital in Phoenix on Sunday, January 17. Williams's father-in-law and brother-in-law—both brain surgeons—did not see much hope for Williams to return to a normal life, and they told Lynda Williams to prepare for the worst.

At the Pasto hospital, the novelty of the volcano victims had worn off, and Andrew Macfarlane and Luis LeMarie were pretty much ignored, so Peter Baxter asked that they be released. Macfarlane and Mike Conway took a commercial flight back to the United States on Saturday, and LeMarie drove with Pete Hall and Patty Mothes back to Quito, Ecuador.

At the Pasto hospital, autopsies were performed on the bodies of José Arles Zapata, Carlos Trujillo, and the three tourists. The cause of death for Zapata and the three tourists was destruction of the skull and exposure of the brain resulting from the impact of falling rocks. Their bodies were covered with lacerations and burns that may have happened after they were initially hit. Carlos Trujillo's corpse was not charred or desiccated like the victim of a fire but cooked, the fat melted and the flesh hardened, leading the pathologist to conclude that he had been caught in a cloud of steam about 200 degrees centigrade. His skull was fractured and he was found 1,500 feet from the crater rim—in two pieces. A falling boulder was the likely cause of the divided corpse.

Although less convinced than ever about the deadly nature of Galeras, the mayor of Pasto and the governor of Nariño took the opportunity to request assistance from President Barco, who visited the area several days later. They hoped he would declare an economic emergency and offer some financial help for the local infrastructure, which they still claimed had been wrecked by the volcano crisis that never was. As in previous bids for help from Bogotá, they were denied.

Ten days after the eruption, on January 24, Goff's team went to the observatory at 4 o'clock in the morning. They had been collecting water samples from springs lower on the mountain, and Goff decided that it was time to go back to the crater. This time, they were asked to sign release forms, freeing the observatory of any liability concerning their well-being if they chose to go into Galeras. That morning and for the following five days, Goff, Gary McMurtry, Tobias Fischer, Andy Adams, and Alfredo Roldán went to the summit of Galeras. They all wore hard hats, carried full-face gas masks, and wore fire-resistant suits. Each day, Galeras would not cooperate. The clouds were so thick and the visibility so poor that Goff chose not to take the team down. Each day, Adams was silently thankful. Finally, on January 30, the day was beautiful and clear, and they began making their way down into the volcano. Bruno Martinelli was with them, but he would stay on the caldera rim and act as the radio link between Goff and the observatory.

A million thoughts raced through Adams's head as he descended into the volcano that had nearly killed him. He was still angry about the way the field trip had been conducted. "I guess our biggest bitch was the lack of safety, and they were observing these seismic events and they were seeing them up to two weeks before." Williams, who had taken a rock in his head, had laughed at Roldán and him in their hard hats. Williams had never tried to get the team to hurry, never exuded any sense of urgency or danger. Sampling in the crater that usually took only an hour or two had taken four. And the *tornillos,* why hadn't he said anything about the *tornillos*—or the paper by John Stix? He knew that Fraser Goff was a stickler for safety, and that they would now take much greater precautions, including checking the seismometers every morning at 4 A.M. Still, Adams could not shake the feeling of dread.

Back in his hotel room, in his black briefcase, he had left a letter written on Hotel Cuellar stationery. On the envelope it read "To my wife, in case of my death." Inside, Adams's letter began: "If you're reading this, I really screwed up."

Alfredo Roldán, on the other hand, believed more in fate: Galeras had spared their lives once, and Roldán considered the steaming volcano more friend than foe. The eruption was not something that would keep him from working in active volcanoes. He had always been extremely careful in the field, wearing his hard hat like a second skull, even in the car on the ride up the mountain. Where he was from, many people brought offerings to the volcanoes. Incense, rum, a candy bar . . .

The weeks following the tragedy were exhausting for Marta Calvache. There were reports and interviews and logistical problems to deal with. Calvache took Fraser Goff's safety plan and implemented it at the observatory. Now everyone who went to the volcano would sign a release form, wear a hard hat, and have an adequate gas mask. They would wear protective clothing and safety boots. The police station would have a radio link to the volcano observatory in Pasto, and all scientists working in the crater would have to submit a safety plan. There were also plans for emergencies.

The consensus was that the eruption of Galeras spit out barely 100 cubic meters of rock—a million times less than the eruption of Mount Saint Helens—and yet it had taken nine people to their graves. There were funerals to attend, but there was little time to mourn. Calvache and Nestor García had lived through so much in the days of Nevado del Ruiz. They had been to its summit just one day before it erupted, had escaped death together by a single day. José Arles Zapata's death left a horrible hole in the lives of all those who worked at the observatory. He had been their shining star—a young scientist, in love with Galeras and undaunted by the political and civil problems of his country.

Back in the United States, Stanley Williams was making a miraculous recovery. It took only three weeks before he was ready to take his story to the news media. His first interview was with the *New York Times*. The article appeared on February 9 and had one particularly strange twist:

> The blast crushed or burned to death six scientists in or near the crater. The lone survivor among the group was Dr. Stanley N. Williams, 40, a volcanologist at Arizona State University who is married and the father of two small children, aged 7 and 4.

Next, from his hospital room, Williams called Bob Tilling, a top-ranking U.S. Geological Survey official in Menlo Park. Williams wanted to know how to get hold of the producers of a television news program. A few days later, on February 12, Stanley Williams appeared—with a shaved, stitched head and a wired jaw—on the NBC *Nightly News* direct from his hospital in Arizona. Shown during a physical therapy session, he tightly gripped a steel walking bar, his face contorted in a grimace as he lifted his broken leg. "There were ten people, and I'm the only one that's alive. It's not an easy thing," Williams told reporter Robert Bazell. "I said, 'Something is going wrong. This is—this is bad news. We've got to get out of here.' And before anybody had a chance to react, it had started—it just exploded, and right in front of me, everybody died in seconds. It didn't just kill them, it completely shattered everybody and—and caused them to be inci . . . burned because their bodies weren't even recoverable."

"And you saw this?" Bazell asked.

"I'm afraid so. It was hard to . . . to see my friends die like that." At the end of the interview, Williams looked into the camera and said, "If . . . if those people lost their lives, and the result is that thousands or millions of people don't die, then I guess it's worth it."

Three days later, on February 15, via a live satellite interview, Katie Couric from NBC's *Today Show* introduced Williams as the only survivor. During the interview, Couric pressed Williams about whether there had been signs that the volcano was going to erupt. He denied it absolutely, claiming that the seismic signals that had preceded the July 16, 1992, eruption *were not present* before the January 14, 1993, eruption.

> COURIC: **Was there absolutely no warning whatsoever that this volcano was going to blow?**
>
> DR. WILLIAMS: It was really amazing. Even in retrospect people cannot look at data and say, "Oh, we should have expected it." We had walkie-talkies, we were listening to people at the observatory, we had every reason to believe the volcano was stable.

> COURIC: In fact, this whole horrifying ordeal really illustrates just how primitive the science of predicting when a volcano is going to erupt is.

> DR. WILLIAMS: Yeah. Well, it's a good example of the frustration. We had an explosion at this volcano last July, and for about a week in anticipation of that explosion there were signs: earthquakes of a certain character, of gasing—a certain change. In January [1993], none of those bad . . . patterns were repeated.

In the weeks and months that followed, Williams went on posing for pictures and proclaiming the need to take such risks in order to save millions of lives. He recounted the chain of events as a hero who had personally battled a monster volcano and been the sole survivor.

It was all very distressing to many people in the scientific community, who thought it was a flagrant grab for fame at the expense of dead colleagues. Most distressing of all, especially to the scientists who had been at the conference, was that Williams was describing himself as the only survivor of a terrible, unforeseen tragedy.

Back at Florida International University, still nursing a sprained ankle and vowing never again to work in an active volcano, Andrew Macfarlane was dumbfounded. He had received a few requests for interviews from *Time* magazine and the *Los Angeles Times*. But the majority of reports were only concerned with Williams, who continued to portray himself as the lone survivor. Bothered by the strange stories, Macfarlane called Williams and asked him about the discrepancy.

"I said, 'Listen, Stan, I want to ask you, why were they calling you the only survivor on the news this morning?' He immediately got defensive and told me that he had been misquoted, taken out of context," says Macfarlane. But the reports continued, and it became clear to Macfarlane that Williams himself was responsible for perpetuating the notion that he alone had survived the blast. There were other false claims as well: Williams said that when he heard the rock slide several minutes prior to the eruption, he told everyone to get out as soon as possible. He also said he had been standing on the crater rim and witnessed everyone die right before his eyes. And again and again, he claimed that there were no signs that the volcano was getting ready to blow.

Williams blamed the tragedy on bad luck. The day of the meeting had been switched because of a power blackout—just a cruel twist of fate. In the end, he had saved lives by downsizing the field trip to the crater. What bothered some members of the scientific community most was that he never showed the slightest bit of remorse, never took any responsibility for the accident, never admitted that mistakes had been made. He was, in fact, the senior scientist in charge, the person who led the field trip into the volcano. Other scientists felt that if they had been in Williams's shoes, overwhelming guilt would have precluded such grandstanding.

Andrew Macfarlane wondered if maybe it was brain damage that was causing Williams to forget about all of the other surviving victims. He found himself in an unenviable and uncomfortable position, feeling like he had to prove that he, Mike Conway, and Luis LeMarie were all Galeras survivors as well. Indeed, they had been even closer to the source of the eruption than Williams had been. Andy Adams, Alfredo Roldán, Carlos Estrada, and Fabio García had been farther from the crater at the time of the eruption, but they were also survivors. Did Williams really believe his strange account?

In the mainstream news, Williams stuck with the story that he was the only survivor, but four months after the eruption, in an article he wrote for the June 1993 issue of *Geotimes,* a scholarly magazine published by the American Geological Institute, Williams told a decidedly different story.

> I got up and tried to run. Only to fall because of the compound break. The others, who were close (Luis Le Marie [*sic*], of Escuela Politecnica Nacional in Ecuador; Andrew Macfarlane, of Florida International University;

and Michael Conway, of Michigan Technological University) and just a short distance down the slope of the volcano, were also injured, but all were able to continue moving down toward the caldera scarp (a much safer distance from the crater).

This short article proved that Stanley Williams did remember other survivors, but it went out to a very limited audience, and Macfarlane worried that the story would never be set straight. He had written to William Broad at the *New York Times,* whose article had called Williams the only survivor, and received no response. As a last resort, he made a call to Christopher Sanders, a colleague of Stanley Williams at Arizona State University. Sanders had begun to feel like he was stuck in the middle of Williams's media maelstrom. Sanders had been at the Galeras workshop, and he knew very well that Williams was not the only survivor. He had always had a bad feeling about Williams and had never fully trusted him, but his constant inconsistencies and bids for fame had become particularly nauseating. He wanted to help Macfarlane, who was understandably upset about the position he was in. "Andy Macfarlane was concerned about Stan's well-being," recalls Sanders. With Williams still in the hospital recovering, Sanders made a plea to Lynda Williams. Sanders let her know that other survivors of the eruption were bothered that her husband was spreading an untrue story, and they were worried that he might have experienced some brain damage. Sanders soon received a memo from Stanley Williams. Brutal and bizarre, it blasted Sanders for interfering and ended with a vicious attack on Mike Conway and Andrew Macfarlane.

When the *New York Times* was published (after all of this) and correctly quoted me as the sole survivor of the scientists working at the crater and the NBC broadcasts occurred, Andy and Mike became very angry for two reasons: a feeling of guilt (for not having done anything for me) and jealousy for the recognition which I received. Today, in *Time* magazine, Andy is quoted as saying that he actually tried to carry me down the volcano. This represents a blatant lie from a guy desperate to cover up the reality of the situation. . . . At this point I am appalled with you and deeply insulted by your repeating this blatant bullshit from two pathetic liars. . . . I hope that we can now put this behind us, so that we can have a future constructive relationship.

Williams then went on a smear campaign. "I started hearing reports," says Macfarlane, "that Stan was spreading these rumors that Mike and I were conjuring up lies to try to ruin him." Mike Conway had recently received his doctorate in the chemistry of volcanoes, and he feared he would be blacklisted in his own small field. Macfarlane, an economic geologist, was less worried, but still wanted to do what he could to remedy the situation.

Volcanologists in the United States and around the world were getting sick of seeing Stanley Williams in the news over and over again. In each report, he said that there was no strange seismicity before the eruption of Galeras. And when addressing safety (which was often brought up by other scientists, but never mentioned in the news reports), Williams claimed that everyone had been given adequate information on the state of the volcano, and that everyone had observed adequate safety practices. About both seismicity and safety, there was one group who knew differently.

Fraser Goff and Andy Adams came back to Los Alamos one month after the eruption, their anger about the botched field trip made worse by the media frenzy surrounding Stanley Williams. Soon after their return, Goff gave a talk to his Los Alamos colleagues lambasting the poor way that the expedition was conducted. It was set up with all the caution of a Sunday afternoon walk in the park, he said, not a trip to an active volcano. He had a set of overhead slides about the trip that began:

Galeras Tragedy
January 14, 1993
Casualties: 9 confirmed dead; 10 injured

Problems
1. No general safety plan provided
2. Poor safety equipment for most
3. Poor radio communication (1 not enough)
4. Poor seismic support (seismologist gone)
5. Essentially no first-aid supplies
6. No tourist restrictions (3 dead)

Comments
1. Poor attitude (?) little respect!
2. Two scientists enter throat of volcano (dead!)
3. Four precursor seismic events unrecognized!
4. A workshop report predicting pending danger is largely
    ignored!

What had happened with the seismic signals was also a mystery. Williams continued to claim that there was no warning, but several weeks after the eruption, Bernard Chouet got his first look at the seismic data from Galeras in the weeks before the January 14 eruption. Meghan Morrissey, knowing that Chouet would be very interested, had brought him the copies of the data upon returning to the United States. Chouet was horrified—and at the same time, strangely vindicated. There had indeed been seismic signals: *long-period* signals. Just like those that appeared in the data before the eruption in July 1992 and exactly what he had forecast in November 1991. Both Stanley Williams and John Stix had been with Chouet in Colombia in 1991 and had been present when he reported his prediction for Galeras. Now Williams was claiming that there had been nothing to worry about in the seismic data, and Stix was backing up his claim. Nine people were dead, totally unnecessarily.

Bernard Chouet was dumbfounded. "What happened at the meeting is the biggest unknown to me," Chouet says. "The 1991

report should have been made available to the people in Pasto."
Chouet had no idea why Williams had missed or ignored the long
seismic signals. Chouet remembers wondering: "Why was I working so hard, if no one was paying attention?"

In the fall of 1993, Stanley Williams, having amazed his physicians,
went back to work at the university. He was still facing reconstruc-
tive surgeries. He wore a hearing aid in his left ear and a metal cage
covered his left leg, encouraging his broken bones to grow back
together. He would prop the gruesome contraption on top of his
desk when lecturing to his volcanology class. Among students, and
to the university in general, Stanley Williams—a professor who had
been known for his dry, monotonous lecture style—had suddenly
achieved celebrity. Stanley Williams, in the eyes of his students, was
cool.

Tobias Fischer and Marta Calvache continued their graduate
work with Williams. Calvache spent most of her time in Pasto, dealing with the day-to-day operations of the observatory and working
on her doctoral thesis on the geology of Galeras. Fischer collected
gas measurements using an instrument that takes remote measurements so that he didn't have to go inside the volcano. This was fortunate, as Galeras erupted two more times, on March 23, 1993, and
on June 7, 1993. In both cases, long-period events were present in
the seismic data before the eruptions. The June explosion completely destroyed the police station. Afterward, all was quiet at the
volcano.

In July 1993, six months after the deadly eruption, Bernard Chouet
received a letter from the journal *Nature*, asking him to "referee" a
scientific paper. (Before a scientific journal will publish a new discovery, the science has to be reviewed by other experts in the same
field.) This particular request, however, was unusual, and the topic
of the paper made Bernard Chouet extremely uncomfortable. The
lead author was Tobias Fischer. Other authors listed included John

Stix, Meghan Morrissey, Diego Gómez, and Roberto Torres—and Stanley Williams. And what made Chouet's stomach turn was that the paper was about long-period siesmicity.

Nature, one of the foremost scientific journals in the world, crosses all scientific disciplines, publishing only cutting-edge science. The discoveries published in *Nature,* which is a British journal, and *Science,* its American counterpart, generate considerable media attention. In this particular paper, Stanley Williams's graduate student Tobias Fischer was comparing long-period seismicity to the levels of sulfur dioxide in the gases coming from Galeras and suggesting the combination's utility in forecasting eruptions.

Chouet was already working on his own paper for *Nature,* which would be a full scientific analysis of his careful study on long-period seismicity and its predictive capabilities. Research on active volcanoes was not like research in a laboratory, where an experiment could be repeated over and over in a matter of days. And because of this, Chouet's work would be the culmination of nine years' worth of groundbreaking research. In addition to detailing his successes at Galeras and at Redoubt Volcano in Alaska, his study would present for the first time (in a mainstream publication) the physical processes that cause a volcano to pressurize and ultimately explode. "Bernard Chouet is the foremost guy in the world. He's the guy who pioneered the approach, describing precisely what's going on in the volcano," says Dave Harlow. It was the greatest advance in the field of eruption prediction in years and was to have been the first time Chouet presented it fully to the scientific community—and the public. Now, Tobias Fischer and Stanley Williams were going to publish before him—about his very own work. The *Nature* editors, aware that Chouet was working on his own paper on the topic, asked him if he would mind reviewing the paper. They knew that there would be a conflict of interest, but there wasn't anyone else they knew who was working in the highly specialized field.

To Chouet, the request to referee the paper was not just a conflict of interest; it was a conflict of his spirit. He could not bear the thought of his work, what he had spent nine years of his life perfecting, being misrepresented. When Chouet read Fischer's paper, he realized that the graduate student had no idea about the science behind long-period seismicity, and Fischer had gotten it all wrong. The paper also implied that the combination of gas and seismic data could be used to predict eruptions. The most annoying thing to Chouet was that the gas data were a totally unreliable and unnecessary part of the forecasting equation. It was the seismic events that were important for predicting eruptions, and that part of the story was 100 percent Chouet's discovery.

Chouet was worried that an erroneous paper on long-period seismicity could do more harm than good, especially when published in *Nature*. As far as he was concerned, there were already nine people dead because his work had been ignored. If he reviewed Fischer's paper, they would actually be publishing in the respected journal *before* him, because Chouet's paper wasn't yet under review. Eventually, he decided that the most important thing was to have the science be right, and he agreed to referee the paper.

Because of the deliberate nature of scientific publishing and editing, it took three edits and six months before Chouet would agree that the science in Fischer's paper was sound and ready for publication. Chouet's own paper was now in review by *Nature*. He was content that Fischer and Williams were grateful for his help, and he did not expect that his work would be eclipsed.

But on March 10, 1994, the day the paper by Fischer was published in *Nature,* Chouet was aghast when he discovered that Stanley Williams was making the news once again—this time, touting *his* discovery on how to predict eruptions.

On March 10, the *Los Angeles Times* ran an article by Kenneth Reich.

> An Arizona State University geologist who was seriously injured last year when Galeras Volcano in Colombia unexpectedly erupted while he and other scientists were studying it, reports today that Galeras and certain other volcanoes may give off a signal before erupting.

If so, according to geologist Stanley N. Williams and five other researchers writing in the British journal *Nature,* it may be possible to predict some eruptions.

And had they possessed the information last year, Williams, 41, and the research party he was part of would certainly have stayed away from the crater on Jan. 14, 1993, when nine people were killed in the eruption, including six scientists. . . .

"We misunderstood the situation," Williams said in an interview this week. "We had fallen into thinking the volcano was stable when it wasn't."

And in a March 10 article, *New York Times* science reporter William Broad covered Williams's "new discovery."

In an interview this week, Williams said the finding was likely to become important in forecasting eruptions. "It's the big step we need for going from theory to a useful tool."

Their report is in *Nature,* the British scientific weekly. The lead author is Tobias Fischer, a graduate student at Arizona State who Williams credited with discovering the method.

The scientists' studies of the two Galeras eruptions last year revealed that each outburst was preceded by faint earth tremors of extremely low frequency and long duration that gradually decreased in length.

The paper by Fischer was based on Chouet's work, but Williams, Fischer, and John Stix were taking all the credit for the discovery.

Ironically, within the scientific community, Williams couldn't claim the discovery himself, because he had said over and over that there were no seismic warnings prior to the January 14 eruption. Admitting that he knew the importance of long-period seismicity would make everyone wonder why he had taken the trip into the crater, especially when the volcano had shown the same signs before its previous eruption. Instead, Tobias Fischer was given the credit. The official story was that it was naive serendipity for the young German—a youthful open mind that had caused him to link the gas and seismic data in order to predict eruptions, something no other scientist had ever thought to do.

In an interview with Williams, the spring 1995 issue of *ASU Research Magazine* told the story of Fischer's discovery.

> Sometimes the best thing a professor can do for an inquiring young scholar is to leave him alone.
>
> That's exactly what happened to Arizona State University graduate student Tobias Fischer last year. On Jan. 14, 1993, a tragic eruption at the Galeras volcano in South America took the lives of six scientists. The eruption severely injured Fischer's mentor, ASU geologist Stanley Williams.
>
> Fischer's largely unsupervised research at Galeras with two fellow ASU students produced a groundbreaking article in the March 10, 1995, issue of *Nature,* one of the world's top scientific journals.
>
> He and co-authors did what no volcanologist had done before: they compared the earthquake data from a volcano to its gas data. They found that monitoring volcanic-gas emissions in connection with seismic events may help forecast future eruptions.
>
> "People who study the seismology of volcanoes don't talk very much to people who study the gases. They don't speak to the people who do the gravity studies. And they don't speak to the geologists, who just look at the rocks," Williams says.
>
> "We get too superspecialized, really. Maybe Tobias was naive enough to not know that he shouldn't cross that boundary. He compared the seismic events beneath the volcano with the gas measurements and he came up with a correlation."

For the volcanology community at large, it was hard to swallow.
Williams had been working on volcanoes since Mount Saint Helens
erupted thirteen years before. He had attended all the conferences,
talked to all of the foremost scientists. Was it possible that he had
completely ignored one of the most important discoveries in the
field at the very volcano that was at the center of his research? It
seemed insane that anyone would take a field trip into the crater of
an active volcano without checking the seismic data. But Williams
did have a reputation as a soloist, seemed hell-bent on predicting
eruptions using gas data, and he had a long history of conflicts with
the U.S. Geological Survey scientists. However hard to fathom,
Williams's story that he had never known there was any seismic
warning of the January 14, 1993, eruption was grudgingly accepted
because there was no one in the United States to contradict him. In
the end, the volcanological community believed that Williams had
either ignored the seismic data, or never looked at it because he just
didn't know any better.

Bernard Chouet, however, still reeling from what he felt was a
blatant hijacking of his work, never fully accepted that Stanley
Williams hadn't known about the long-period seismic data before
the eruption that killed those he called his friends.

Chouet's paper was still in review at *Nature,* and he fervently
hoped that he could ultimately get the recognition he deserved
for his discovery. When his work came back from the editors at
the prestigious journal, he was dumbfounded. The same editor
of *Nature* who had asked him to referee Tobias Fischer's paper
three times asked him to remove any mention of Galeras in his own
paper. Her reasoning: *Nature* just published a paper on Galeras—by
Fischer and others. Incredibly, the editor also requested that Chouet
downplay his discovery's usefulness for predicting eruptions.

The Swiss scientist was appalled. He knew that *Nature* wanted
to stay out of any controversy, and he knew he had put a loaded line
into his report when he mentioned "the tragic failure to recognize
the significance of precursory long-period seismicity at Galeras."
Now she was asking him to delete any references to Galeras, and to

use Fischer's paper as a reference for all of the work done on the Colombian volcano. Nine years of work on a significant discovery had been eclipsed by Tobias Fischer, a 25-year-old graduate student, and Stanley Williams, who claimed to know nothing about long-period seismicity. And Chouet had completely rewritten the long-period seismicity part of the paper for them. "I felt hit by a double whammy. I felt like, hey, first of all, don't referee any more papers for these guys. And it really offended me because of the succession of events." Furious, but trying to maintain his composure, he wrote this letter to *Nature*.

> In light of comments concerning Galeras and the application
> of the LP (long-period) forecasting strategy there, I would like
> to seize upon the opportunity of this letter to give you some
> background. This is, perhaps, the best example of the potential
> forecasting power of shallow LP activity. In November 1991, I
> was called upon to visit Galeras and offer advice as to the
> future eruptive potential of this volcano. At the close of this
> trip I wrote a report which stated scenarios for future activity
> drawing attention to the extreme importance of tracking LP
> events, especially if those turned out to be deeper than the
> surficial activity that was observed then (Fig. 3 of my paper
> gives an example of this surficial activity). This report was
> distributed to the Observatory of Pasto, the director of
> Ingeominas in Bogotá, as well as other agencies interested in
> the Galeras situation. Although the report specifically included
> as one of the likely scenarios the LP activity that was to lead to
> the eruptions of 1992 and 1993, no one apparently paid
> attention to it. In January 1993, the danger presented by the
> LP activity that was seen then was known. That is why I talk
> in my paper about "the tragic failure to recognize the
> significance of precursory long-period seismicity at Galeras."
> There was no need for people to die on this mountain given
> the information available at the time, but of course this is a
> very difficult and sensitive point to make. Then, as you will

recall, you sent me the paper by Fischer and others to review. Not only did I coach Tobias Fischer and Meghan Morrissey personally during a trip to ASU, I also spent time helping them to develop a model they had little feel for at first (my last letter to you dated January 7 alludes to that effort). Now, I read in the *New York Times, Science News,* and *News and Views in Nature* that these people are predicting eruptions, and at the same time I am told that I should not emphasize a side of the story on which these news are based. I found this a bit hard to swallow. This being said, however, given the obvious sensitivities of people concerning what happened at Galeras, it is not worth the turmoil and emotion that it will generate to focus on Galeras in this article and I would be happy to drop that story entirely. In fact, cutting it out will not detract from the central purpose of informing the scientific community that seismology is now yielding quantitative clues as to the physical processes and conditions inside a volcano that can provide the basis for building a forecasting strategy based on physical understanding rather than empirism or pattern recognition. . . .

But in the end, he decided he couldn't publish his work if it was going to be censored. He was incensed with Williams and with *Nature.* He felt like the system had steamrolled him. He wanted to fight, to argue, to challenge the editor, to blast Fischer and Williams, but the anger was beginning to destroy him. "I complained bitterly and then put the paper on the shelf and decided that I wouldn't have anything to do with it. For two and a half years it remained on the shelf," he says.

What never surfaced in the ongoing debate about what Stanley Williams did or didn't know about the seismic data before the deadly eruption was a document Williams had written just over one year earlier. It was a grant proposal awarded in December 1991—just one month *after* Bernard Chouet filed his original report in Pasto (at a meeting attended by both Williams and John Stix) that had provided such an accurate road map for seismicity at Galeras.

Galeras Volcano 1991 Eruption Studies
NSF Award Abstract: #9202195
Latest Amendment Date: December 24, 1991
Award Number: 9202195
Award Instr: Standard Grant
Prgm Manager: Maryellen Cameron
Start Date: February 1, 1992
Expires: July 31, 1993 (Estimated)
Expected Total Amt.: $8,200 (Estimated)
Investigator: Stanley N. Williams
Sponsor: Arizona State University

Abstract
In November 1991 Galeras Volcano in Colombia showed
numerous signs of impending eruption, very similar to those of
the neighboring volcano Ruiz just before it erupted a few years
ago with a loss of over 20,000 lives. This grant will enable the
PI (principle investigator) and his student to go to Colombia
before any eruption comes and establish a small seismic
network on the flanks of the volcano. They will then try to
correlate seismic tremors with specific gas emissions as
determined by COSPEC (correlation spectrometer). This
information about the relationship between seismicity and gas
emission could be very important in attempts [to] predict
future eruptions of similar kinds of volcanoes. If no eruption
occurs while the PIs are there, local geologists will use the
equipment to make measurements when and if it does occur.

The grant proposal Williams had submitted to the National
Science Foundation in his own name would turn out to be the study
published in *Nature* that Williams claimed had been created and
conceived principally by his 25-year-old graduate student, Tobias
Fischer, without Williams's guidance or knowledge. The grant proposal abstract, written over a year before the Galeras tragedy,
proved that Williams was indeed aware of the importance of seismic
data for forecasting eruptions at Galeras, but it was never brought

to light. In the years that followed, Williams would continue to claim that on January 14, 1993, Galeras had been as quiet as an active volcano can be, and that he had been aware of no warnings to the contrary when he led the ill-fated trip to Galeras's crater.

This was the story he told to the newspapers. It was the story he proclaimed in television interviews, the story he told the volcanologists from the U.S. Geological Survey, the story he entertained classes with at Arizona State University—and the story he told to the widows and children of the men who died.

Most important, this was the way he presented the field trip to his colleagues José Arles Zapata, Geoff Brown, Carlos Trujillo, Fernando Cuenca, Igor Menyailov, and Nestor García as they planned what would be their final descent into Galeras: that the trip to the volcano's crater would be "a piece of cake"; that, based on all the available data and his own considerable experience and expertise, there was absolutely no cause for concern, nothing to fear.

To the people of Colombia, the only true volcanic tragedy in their country took place in Chinchiná and Armero on the night of November 13, 1985. It was a tragedy of God and man, of beliefs and culture, of ego and ignorance, and it was a catastrophe that obliterated precious land and property and buried thousands of their countrymen. In 1993, when six scientists died at Galeras, it was not considered a tragedy, but a ridiculous irony. While the volcanological community mourned, the Colombians laughed. The very experts who should have known better walked right into the mouth of a volcano. Half of them never returned.

EPILOGUE

I FIRST MET STANLEY WILLIAMS in January 1996 in the small, grungy auditorium in the Earth Sciences building at UC Riverside, where I was working on my master's degree. Williams had come to speak to the geology department, and we'd been anticipating his arrival all day: to geologists, nothing is more exciting than an active volcano, so we tend to idolize anyone with the guts to take one on. We couldn't wait to meet the man who had become famous as the sole survivor of a volcanic tragedy.

He was not what I expected; he looked more like a computer programmer than somebody who'd muscled through a volcanic firestorm. He stood at the front of the room and told the story of Galeras, of how he'd watched his good friends blown to a million pieces in front of his eyes, how he had lain helpless until his own female graduate student carried him out. And most interesting of all, he told how, after the eruption, another of his graduate students discovered that the Colombian seismologists had seen some threatening signals in the seismic data right before Galeras erupted. If only we had known, he said. If only scientists from different disciplines would talk to each other; if only the gas guys had talked to the seismic guys. In the end, he seemed to say, it was a simple lack of communication that had cost nine people their lives.

Williams had come to UC Riverside at my graduate adviser's invitation, and I was asked to the luncheon afterward. He had said in his talk that he was already back at it—tackling active volcanoes all over Latin America. I asked his wife, Lynda, what that was like for her and their kids. She shrugged and sidestepped the question, seeming to be resigned to the fact that she wasn't going to be able to keep him from what he seemed determined to do.

I remember thinking that the Galeras tragedy was a great story, and that Williams should write a book about it. In the meantime, I went on with graduate school and found my own volcano—Mount Rainier—to work on in 1997.

Two years later, Stanley Williams came back into my life. A book agent called me after he heard Williams give a presentation and asked me if I knew the story of Galeras. I did, I told him, a fascinating tragedy. It seemed that Williams was finally ready to tell his story, and I was dying to be the person to help him write it. A few months later, I got word that he already had a coauthor, had sold his book proposal to a New York publisher for a big sum of money, and was well on the way to writing the next adventure-tragedy bestseller.

I'd missed my chance to write the great volcano story, a fact I lamented over drinks with a University of Hawaii volcanologist. When I brought up Williams's good fortune, he stiffened visibly. "You know," he said after a moment, "there's a bigger story there." He was guarded, and I thought I detected a note of bitterness in his voice—professional jealousy, perhaps? But he got me thinking that maybe there *was* a bigger story—not just about Stanley Williams but about the whole group of scientists who died on Galeras, and about what drives seemingly rational scientists to risk their lives.

As we finished our beer and got up to leave, the volcanologist said one thing more about Stanley Williams: "A lot of people have a problem with Stan's safety practices."

In the library at NASA's Goddard Space Flight Center, I found a special volume dedicated solely to the Galeras eruption, published by the *Journal of Volcanology and Geothermal Research* in 1997. Toward the end of the volume was an article by Peter Baxter, M.D. He had been at the Galeras workshop, and his paper was a compilation of interviews from the Galeras survivors.

Survivors? I'd always heard that Williams *alone* had survived Galeras. And yet here was testimony from five other scientists who'd *also* survived the eruption in January 1993.

This was the first in a long string of discrepancies I discovered in the story Stanley Williams told to the media. My investigation—the result of which is this book—led me to Colombia and eventually to interview everyone still living who had played a part in the tragedy.

But in the end there was one last survivor I needed to talk to. It was a 109-degree June day in Tempe when I arrived at Arizona State University. Williams was teaching a summer school class, and I wanted to give him a chance to tell his side of the story.

He was lecturing in a big auditorium, and a friend and I stood in the back and listened. He looked as I'd remembered him and spoke in the same monotone. He wore shorts; the scars on his legs were clearly visible. A plastic hearing aid filled his left ear.

At the end of class I walked up and introduced myself. I didn't expect a warm welcome—by now I had my own book contract, as he was well aware—but we shook hands and made small talk on the way up to his sixth-floor office. I was nervous and I could tell he was, too, but he seemed willing to answer my questions. I asked him about the days after the Armero tragedy, and how he came to work on Galeras. He knew I was a scientist, but he gave me the same sound-bite answers that I'd read in so many news clips and television transcripts. For the most part, he was cordial—until I asked specifically about the January 14, 1993, eruption of Galeras.

He shifted in his seat, picked at his fingernails, then politely told me that it was time for me to leave. But I had one more question.

"What about the seismicity? The night before the eruption, the seismologists said that you all had discussed this and they were worried."

"Who said that?" he demanded.

I listed my sources for Williams. They included Fernando Gil, whom Williams had known since 1988, when he had asked to include Gil and Bernard Chouet's work on long-period seismicity in a special Nevado del Ruiz journal that Williams was editing.

Williams grew increasingly uncomfortable. It seemed he had never expected this line of questioning.

"I wasn't aware of anything like that. Besides, I'm not a seismologist."

Then he told me he was through talking to me, so I closed my notebook and shook his hand. Despite the fact that I had to root out the truth of what happened at Galeras, and I already knew where this contentious story was going, I still felt sorry for Stanley Williams, for his suffering and the trauma he'd endured. On a volcano in Colombia, this man and fifteen others had been caught in a cataclysm. Six people—including the leader of the field trip, Stanley Williams—had survived. Nine others had not.

As we said good-bye I felt that those facts were as much a part of his daily experience as his hearing aid and the scars on his body—all pointing back to a day of unmitigated tragedy.

ACKNOWLEDGMENTS

THIS BOOK BELONGS to the people who lived through the tragedies of both Nevado del Ruiz and Galeras. This is a Colombian story, and without the candor and generosity of the Colombian scientists involved, it would have been completely impossible to uncover the complicated story behind the deaths of many thousands in November 1985 and nine others in 1993. To all those who shared their stories, unveiled their emotions, detailed their adventures, I am incredibly grateful.

In Bogotá, at the National Institute of Geology and Mines, Juan Duarte, Fabio García, Jaime Romero, Fernando Muñoz, Hector Cepeda, and Roberto Torres recounted memories from the decade between 1984 and 1994. Yesenia Arzuza, the gracious Ingeominas public-affairs officer, pinned down scientists with crazy schedules, set up interviews, and sent me near-real-time e-mail answers to dozens of questions.

At the Instituto Geofísico de Los Andes Colombiano, Tito

Vargas and Oscar Rodriguez took me back to the earliest days of Colombian seismology, the first seismometer in the Bogotá foothills, and the life of Padre Jesús Emilio Ramírez. Luis Eduardo Peñuela negotiated the suicidal streets of Bogotá, the highway to Laguna de Guatavita, and except for that first night, always remembered to pick me up at the airport.

In Manizales, Carlos Arturo Garzón, director of the volcano observatory, was an invaluable resource. He kindly gave me complete access to the scientists at the observatory, organized a trip to the volcano, and hooked me up with a military helicopter to take me across the cordillera. Gloria Patricia Cortés and Hector Mora shared painful memories of the days after Armero and the day that Galeras erupted. Gustavo Garzón—who was not part of Ingeominas for either of the tragedies—became an invaluable resource to me nonetheless, and for the indigenous Colombians' take on their volcanoes, among other details, I have Gustavo to thank. On a beautiful March day, I joined Nevado del Ruiz's two most renowned tour guides, Fernando Gil and Maria Luisa Monsalve, who took me to the great mountain and shared its secrets. In the city, Pablo Medina's wife, Clarita, and daughter, Lucita, heaped thick stacks of references on me that Pablo had meticulously archived during the crisis of Nevado del Ruiz. Pablo recounted the tragic year by phone and e-mail from his job two hours outside of Manizales across a guerrilla-controlled highway. At the University of Caldas, Adela Londoño recounted fond memories and showed me beautiful photographs of her days working with Nestor García and Marta Calvache before the eruption of Nevado del Ruiz. Maria Elena Vivas told the tale of the rattling, freezing *refugio* from her job at C.M., Manizales's swank private club. Omar Darío Cardona, on holiday in Manizales to visit his parents, took Norm Banks and me up to a ridge to look over his beautiful city and recounted the pain of the Ruiz tragedy and the debacle of Galeras. I must also applaud Omar for his detailed account of the Galeras crisis in the *Journal of Volcanology and Geothermal Research,* 1997. For the price of a tank of gas, Fabio Diaz piloted a Hughes 500S 16,000 feet over

Chinchiná so that I could inspect the mudflow damage that—fifteen years later—still scars the channel 100 feet above the river's bank.

In Pasto, at the volcano observatory, Marta Calvache and Bruno Martinelli recalled their own stories and then kindly enabled me to interview all of the scientists and support personnel who lived through and lost friends to the Galeras tragedy: Adriana Ortega, Ricardo Villota, Diego Gómez, Carlos Estrada, Milton Ordoñez. On a cool night in Pasto I shared volcano stories and vodka with Natalie Ortiz and Gustavo Córdoba. Jose Cuellar, the owner of Hotel Cuellar, bought Alfredo and me dinner and offered hours of insight into the *Pastusos'* take on Galeras and the imagined volcano crisis. Luis Ruales, our waiter, remembered the week that the scientists came to Pasto and dropped something into the volcano. Winston Virachaca and Eduardo Cruz, the journalist and cameraman who recorded the Galeras tragedy, recalled the day—January 16, 1993—as their last trip ever to Galeras.

In Quito, Ecuador, Pete Hall detailed how the Nevado del Ruiz crisis played out in the press and between the battling science and government factions, and his wife, Patty Mothes, gave up hours and hours of her precious time to help me understand the complicated crises. (In return, Pete made me promise to pack up a car battery to 20,000 feet next time I'm in town.) Luis LeMarie has left Quito and now lives and works in the Netherlands, but he graciously recollected in writing his memories of the Galeras eruption for this book.

I want to give special thanks to Alfredo Roldán, who guided me to Pasto and Galeras, and Norm Banks, whose stellar reputation with the Colombian scientists opened doors from Bogotá to Manizales to Pasto.

Fraser Goff and Andy Adams from Los Alamos National Lab first made me aware of the strange inconsistencies in the stories that came out of Colombia after the 1993 Galeras workshop. Andrew Macfarlane, from Florida International University, and Mike Conway, from Arizona Western College, recounted not so much their harrowing tale of survival but the incredibly frustrating years since, as they watched Stanley Williams portray himself as the only survivor of Galeras. Bernard Chouet revealed the story of his discovery, and the pain and frustration of having his work ignored—resulting in the deaths of nine people. Other North American scientists who attended the Galeras workshop and offered me their recollections include Steve McNutt, University of Alaska; Meghan Morrissey, USGS; Chuck Wood, University of South Dakota; Christopher Sanders, Microsoft; Greg Arehart, University of Nevada, Reno; Tobias Fischer, University of New Mexico; and John Stix, McGill University, Montreal, Canada.

Within the U.S. Geological Survey, John Ewert, Dan Miller, Randy White, and Dave Harlow recounted their Colombia experiences. I must give a very special acknowledgment to Jack Lockwood, USGS emeritus, for recording the harrowing experience of Juan José Restrepo, among other Armero survivors, and also to Juan José Restrepo for his courage to tell the horrifying tale of living through the avalanche. Peter Mouginis-Mark, from the University of Hawaii, initially gave me the idea to look for a story deeper than the one reported in the media. Barry Voight, Penn State University, wrote a wonderfully detailed report on Nevado del Ruiz in the *Journal of Volcanology and Geothermal Research,* 1990.

Others who offered me background on the science, the people, and the politics of both tragedies include Barney Berger, USGS; Tom Aldridge, Carnegie Institute of Washington, emeritus; Claudia Zapata, ASU; Rick Wessles, ASU; Jimmy Hinacapie, University of El Paso; and John Tomblin, emeritus director, UN Disaster Relief office.

Dave Zimbelman, an incredible athlete and scientist, first introduced me to Mount Rainier and taught me how to stay safe on the 14,000-foot pile of rubble. George Breit, USGS, was always around to answer questions such as "How can cadmium kill you?" Arlene Collins offered me her beautiful photographs of the Colombian scientists working in the aftermath of the Armero tragedy. Patrick O'Neill, *Oregonian* reporter, tracked down resources for me while I was on the road, and Fred Lovinger, *Oregonian* retired copyeditor, offered endless support. The faculty and staff of the Earth Sciences

Department of the University of California, Riverside, gave me a thorough background in field geology and taught me to use the scientific method, which, it turns out, is much like journalism. For hours of help translating difficult science terms, a special thank-you to Margarita Martí, from Isla Margarita, Venezuela.

An invaluable resource for getting around Colombia is the Lonely Planet Travel Guide, 1995 edition, with a couple of updates circa 1997. Things have changed a lot, but it's still the best reference. Two other notable book references are *The Making of Modern Colombia: A Nation in Spite of Itself* by David Bushnell, and *Historia de Los Terremotos en Colombia* by Jesús Emilio Ramírez.

Emilie Lorditch, a wonderful friend and a great researcher, found me the most obscure details at a moment's notice, and Chris Oliver did an assortment of favors for me that made my life easier this past year. Sterling Spangler, who is like a brother to me, created beautiful maps of Colombia, Galeras, and Nevado del Ruiz. My dear friend Nancy Austin offered her support when I went to interview Stanley Williams, and Melinda Wright dug up her meticulous notes that helped me keep my dates straight.

There are special places in my heart for Holly Hodder, who walked into my life one day like a hurricane and changed my future forever; my unstoppable agent, coach, and grounding medium Peter McGuigan; Peter's ultra high-energy assistant and research junkie Amy Lotven; Nikola Scott, the incredibly competent, creative, and reassuring assistant to my editor; Dan Conaway, my wonderfully good-natured editor at HarperCollins, who worked extraordinarily hard for me with never-ending enthusiasm for this story; and my grandmother, who always told me I would write a book someday. Above all, my love to Mom, Dad, and Tim. There are no words to express how much your support and encouragement mean to me.

INDEX

Entries in *italics* refer to maps.